Algorithms for PID Controller 2019

Algorithms for PID Controller 2019

Guest Editor

Anastasios Dounis

MDPI

Basel • Beijing • Wuhan • Barcelona • Belgrade • Novi Sad • Cluj • Manchester

Guest Editor
Anastasios Dounis
Department of Biomedical
Engineering
University of West Attica
Athens
Greece

Editorial Office
MDPI AG
Grosspeteranlage 5
4052 Basel, Switzerland

This is a reprint of the Special Issue, published open access by the journal *Algorithms* (ISSN 1999-4893), freely accessible at: https://www.mdpi.com/journal/algorithms/special_issues/PID_Controller.

For citation purposes, cite each article independently as indicated on the article page online and as indicated below:

Lastname, A.A.; Lastname, B.B. Article Title. *Journal Name* **Year**, *Volume Number*, Page Range.

ISBN 978-3-7258-4815-7 (Hbk)
ISBN 978-3-7258-4816-4 (PDF)
https://doi.org/10.3390/books978-3-7258-4816-4

Contents

About the Editor

Anastasios Dounis

Anastasios Dounis received his B.Sc degree in physics (1983) from the Department of Physics, University of Patras, Greece, his M.Sc. degree in electronic automation (1988) from the Department of Physics, National and Kapodistrian University of Athens, Greece, and his Ph.D. degree (1993) entitled "Expert Fuzzy of Logic System for Thermal and Visual Comfort Control in Buildings" from the School of Electrical and Computer Engineering, Technical University of Crete, Greece. He is a Professor in the Department of Biomedical Engineering at the University of West Attica, Greece. He is the director of the Artificial-Computational Intelligence of Multidisciplinary Applications Lab (ACIMA Lab) at the University of West Attica. His research encompasses many aspects of Computational Intelligence, Intelligent Control, Fuzzy Systems, Neural Networks, Evolutionary Algorithms, Extensions, Fuzzy Sets, Multi-Agent Systems, Machine Learning, Deep Learning, Time Series Prediction, Smart Energy Management in Buildings/Hospitals, Micro-grids, Fuzzy Multi-Attribute Group Decision-making, Intelligent Medical Diagnosis, and Medical Image Classification. He has participated as a Guest Editor in seven special editions of internationally recognized journals and as a reviewer in more than 30 journals. He is the author or co-author of over 100 published papers in international journals and conferences. He is a member of the Editorial Boards of eight Journals. For three continuous years, encompassing 2021-2023, Prof. Dounis was included in the World's Top 2% of Scientists List at Stanford University ranking, in the field of "Energy" and in the subfield of "Artificial Intelligence & Image Processing", which is among his main scientific fields of expertise. (Elsevier Data Repository, V6, doi: 10.17632/btchxktzyw.6).

Preface

Conventional PID (Proportional–Integral–Derivative) controllers are most widely used in industrial applications because of their simple, robust, economically efficient, and good performances. To date, PID control performance remains limited. The requirements for control precision have become higher, and the real systems have become more complex, including higher-order, time-delayed linear, and nonlinear systems, and systems without a mathematical model and uncertainties. The goal of control algorithms is to determine the optimal PID controller parameters. Practically all PID controllers made today are based on microprocessors. This has created opportunities to provide additional features, such as automatic tuning, gain scheduling, and continuous adaptation. In addition to the conventional approaches such as the Lyapunov approach and PID control system analysis, there are more advanced and intelligent algorithms for PID tuning methods and metaheuristic algorithms, such as the Genetic Algorithm, Particle Swarm Optimization, Ant Colony Optimization, Big Bang–Big Crunch, etc. In addition, sophisticated control strategies, such as predictive control, self-tuning methods, and fuzzy and neural algorithms, are designed to overcome the problems associated with the regulation of PID controller gains.

Anastasios Dounis
Guest Editor

algorithms

MDPI

Article

Quadratic Model-Based Dynamically Updated PID Control of CSTR System with Varying Parameters †

Dushko Stavrov *, Gorjan Nadzinski, Stojche Deskovski and Mile Stankovski

Faculty of Electrical Engineering and Information Technology, Ss. Cyril and Methodius University in Skopje, Rugjer Boshkovic br. 19, 1000 Skopje, North Macedonia; gorjan@feit.ukim.edu.mk (G.N.); stojce.deskovski@gmail.com (S.D.); milestk@feit.ukim.edu.mk (M.S.)

* Correspondence: dushko.stavrov@feit.ukim.edu.mk

† This paper is an extended version of our paper published in 14th ETAI conference (ETAI 2018).

Abstract: In this paper, we discuss an improved version of the conventional PID (Proportional–Integral–Derivative) controller, the Dynamically Updated PID (DUPID) controller. The DUPID is a control solution which preserves the advantages of the PID controller and tends to improve them by introducing a quadratic error model in the PID control structure. The quadratic error model is constructed over a window of past error points. The objective is to use the model to give the conventional PID controller the awareness needed to battle the effects caused by the variation of the parameters. The quality of the predictions that the model is able to deliver depends on the appropriate selection of data used for its construction. In this regard, the paper discusses two algorithms, named 1D (one dimensional) and 2D (two dimensional) DUPID. Appropriate to their names, the former selects data based on one coordinate, whereas the latter selects the data based on two coordinates. Both these versions of the DUPID controller are compared to the conventional PID controller with respect to their capabilities of controlling a Continuous Stirred Tank Reactor (CSTR) system with varying parameters in three different scenarios. As a quantifying measure of the control performance, the integral of absolute error (IAE) metric is used. The results from the performed simulations indicated that the two versions of the DUPID controller improved the control performance of the conventional PID controller in all scenarios.

Keywords: dynamically updated PID; conventional PID controller; continuous stirred tank reactor (CSTR) system; varying parameters; quadratic error model

1. Introduction

To date, the PID (Proportional–Integral–Derivative) controller is the most widely used controller in industry. Certain sources indicate that it is used in up to ∼90% of the industrial plants at low control level [1,2], mainly because it is a simple, robust and cheap controller that offers good control performance in most applications. The most challenging task in PID control design is the determination of its coefficients [3,4]. Moreover, real plants are subjected to change in their parameters, i.e., parameter drift, as they operate. Hence, if the PID controller was initially designed to work for a particular operating point, after the parameters drift, the controller should be adapted to the new operating conditions. If there is no supervisory system present that automatically ensures that adaptation, one should track the parameters drift and occasionally retune the PID parameters.

To tackle the problem of changing plant parameters, adaptive control schemes are used. Among the different adaptive control schemes discussed in the literature, the most widely used is the model reference adaptive control (MRAC). This algorithm adapts the controller's parameters with the aim to force the plant to behave as some given ideal model [5,6]. The adaptive control schemes fall into two categories [3], (i) indirect methods [5,6], and (ii) direct methods [7–9]. In the former, at first, a model of the plant being controlled is identified, and subsequently the obtained model is used to adjust the parameters of the PID controller. In the direct methods, the estimation of the controller parameters is carried out online so that the error between the reference model output and plant output tends to be zero. In general, these methods face difficulties emanating from two sources; one of the difficulties arises from the fact that the controlled plants are black boxes and their identification is a tough and complex task, and the second one stems from the fact that it is difficult to guarantee that the controller parameters will converge to the desired values, and achieve the goal of zero error.

Another common approach for countering the consequences of varying plant parameters found in industry is the robust control approach [10]. The ideas emerging from the development of the robust control paradigm were specialized for the case of the PID controllers, resulting in numerous tuning methods for robust PID control [11]. One of the most often used is the internal-model-based (IMC) PID robust tuning method [11–13]. A novel robust approach is presented in [14], to deal with the negative implications caused by the inherent feature of the IMC tuning of pole-zero cancelation. Lately, the robust control community has shifted its attention toward tuning methods based on multi-objective optimization [11,15,16]. The tuning parameters in these methods are obtained as a solution to an optimization problem which is often non-convex. The discussed robust tuning methods have proven effective in improving the robustness in a variety of control systems. However, they have certain downsides, and some of them are pinpointed as follows: firstly, in the tuning process it is necessary to account for all the parameters that can possibly vary and the bounds of their variation in advance; secondly, they usually need one or multiple linearized models of the nonlinear plant to design the controller; lastly, the methods based on optimization techniques come with a high computational burden as the optimization problem therein is non-convex.

To tackle some of the aforementioned problems, in this paper we discuss a simple and computationally non-intensive, hands-on control algorithm which is an upgraded version of the conventional PID controller—a dynamically updated PID (DUPID) control scheme [17]. This control approach uses a local quadratic model [18,19], of the plant control error to improve the robustness of the conventional PID against the change of the plant parameters. Based on this model the PID control value is adapted directly without retuning its coefficients, to steer the plant output in close proximity of the set point. The DUPID controller is mainly meant to work well for plants found in the process industry. The plants encountered there are often monotone (inert), stable and primarily affected by slow and gradual degradation in their physical components over time. The DUPID control scheme is made of two parts: a PID controller and a supervisory mechanism (SM). The SM plays the role of the adaptation mechanism; it detects the change in plant parameters by locally modelling the recent history of the control errors and uses that model to produce an incremental value that is added to the PID control value.

The control performance of the DUPID control scheme is dependent on the quality of the predictions the error model can deliver. To establish a locally precise error model, we must ensure that the data used for modelling meet certain criteria [18]. In this regard, this paper discusses two versions of the DUPID algorithm, the 1D DUPID and the 2D DUPID. The effectiveness of both versions of the DUPID controller was compared against the conventional PID controller, when each controller was applied to control the CSTR system with varying parameters [20]. We assumed that the varying parameters in question are: the heat transfer coefficient h and the feed fluid temperature T_f. Based on this assumption we defined three different scenarios in which these parameters vary linearly over time. The observed control performance of each controller was quantified and compared against the other controllers based on the IAE metric values.

The rest of the paper is structured as follows: Firstly, the mathematical background of the PID and DUPID are presented, and the DUPID and its two versions are discussed in more detail. Secondly, the CSTR system with varying parameters is analyzed. Thirdly, the PID controller and both versions of the DUPID controller are tested when applied to the CSTR system. Finally, conclusions and an outlook for future work are given.

2. Methods

2.1. The PID Controller

In industry, more than ten particular forms of PID realizations can be found [4,21]. In this paper, the parallel PID realization is used. The mathematical definition of parallel PID realization is given with Equation (1). As it can be observed, the control value is produced in one step and it depends on the current error as well as on previous control errors. The control error in the current iteration $e^{(i)}$ is defined as the difference between the current values of the set-point (SP) and the process value (PV), $e^{(i)} = SP^{(i)} - PV^{(i)}$. The PID tuning parameters are K_p, K_i and K_d, each corresponding to the P, I and D terms respectively. Finally, u_0 is the bias in the control signal and T_i is the simulation time step:

$$u_{PID}^{(i)} = u_0 + K_p e^{(i)} + K_i \sum_{m=1}^{i} e^{(m)} T_i + K_d \frac{e^{(i)} - e^{(i-1)}}{T_i}. \tag{1}$$

In Figure 1, the feedback control loop consisting of a PID controller and a control object (CO) is given.

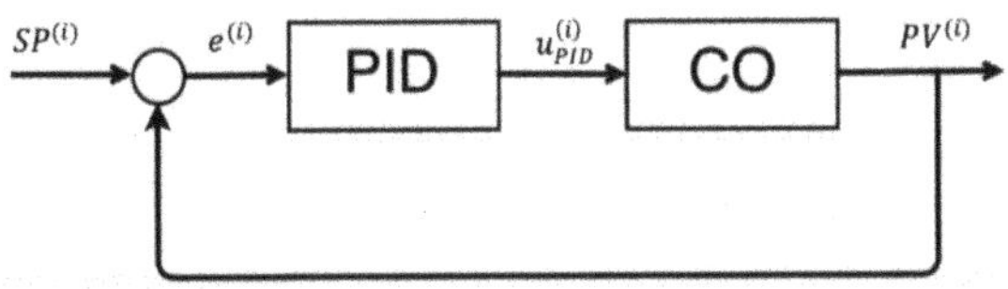

Figure 1. PID (Proportional–Integral–Derivative) control loop structure.

2.2. The DUPID Control Scheme

The DUPID is composed of two main elements: (i) a PID controller and (ii) a supervisory mechanism (SM) (see Figure 2, dashed rectangular). The former acts as a mechanism for preserving CO stability, while the latter is responsible for calculating the incremental value $\Delta u^{(j)}$ in the PID control value. The value $\Delta u^{(j)}$ plays the role of an additional bias when the plant parameters are gradually drifting and the integral action produced by the PID controller is not enough to counter the resulting controller performance degradation caused by the varying parameters. The DUPID controller tracks the slow variation of the parameters and calculates $\Delta u^{(j)}$ which helps the PID controller to steer the plant in the direction of smaller control error.

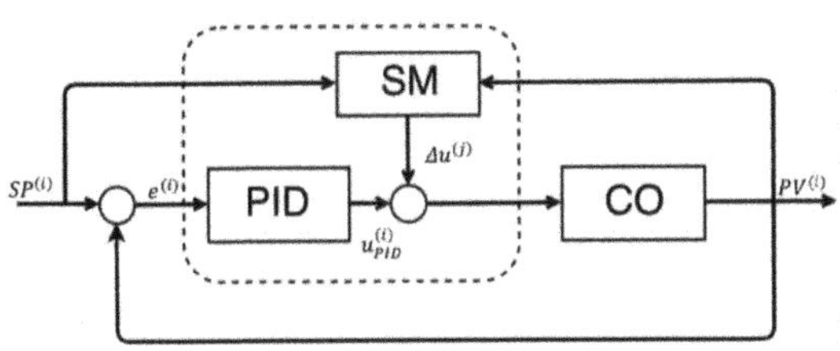

Figure 2. Dynamically updated PID (DUPID) control loop structure.

The DUPID produces the control value in two phases: the first phase is completed after the PID control value $u_{PID}^{(i)}$ is calculated and the second is completed after $\Delta u^{(j)}$ is obtained by the SM.

With the conclusion of the second phase the aggregate control value $u_A^{(i)}$ is applied to CO, Figure 2.

$$u_A^{(i)} = u_{PID}^{(i)} + \Delta u^{(j)} \tag{2}$$

The term $\Delta u^{(j)}$ is generated by the SM at specific moments in time j, defined by the values found in the vector DV. The values of DV are set a priori by the operator and are constant during the plant operation. The role of DV is twofold: its values define the number of simulation time steps that should pass before a regression point is acquired, and at the same time its values define the frequency of $\Delta u^{(j)}$ calculation. The fundamental component of the SM is the quadratic error model given with Equation (3). The quadratic model observed here is the simplest one dimensional (one input) quadratic form:

$$E_M\left(\Delta u^{(j)}, p\right) = A\left(\Delta u^{(j)}\right)^2 + B\left(\Delta u^{(j)}\right) + C, \tag{3}$$

where, $p = \{A, B, C\}$ is the vector of unknown parameters and E_M is the expected control error (the dependent variable). This model locally describes the functional relation between the $\Delta u^{(j)}$ and the plant control error. As new plant data are being gathered, this model is iteratively updated based on the last N_{pp} point pairs $\left(\Delta u^{(j)}, e_s^{(j)}\right)$, where $e_s^{(j)}$ is assigned to the i-th control error point $e^{(i)}$ attained at moment j. The objective is to use this model in iterative fashion to estimate a sequence of $\Delta u^{(j)}$ values for which the error caused by the parameters drift will be gradually decreased. To achieve this, we assume $\Delta u^{(j)}$ to be the root of the model given with Equation (3). If the roots are complex numbers, then the incremental value is calculated by differentiating Equation (3), $dE_M(\cdot)/d(\Delta u) = 0$. Usually, a minimum of three points is required to obtain the parameters p [22]. The model based solely on the minimum required number of points has proven to be ineffective, especially when the regression data are improperly distributed or contaminated with noise [11]. These downsides can be dealt with, by choosing the number of regression points to be a compromise between the minimum required number of points and a certain threshold $\left(N_{pp}\right)$. The initial model resides on the first N_{pp} points, which are gathered consecutively in N_{pp} simulation time steps, as the plant is operated. Since $\Delta u^{(j)}$ does not exist at first, informative prior data are needed to establish the first model. Using a non-informative (poor) prior data can cause the DUPID to fail to converge. Therefore, the first N_{pp} incremental values are calculated by Equation (4), which is a simple metric that adds extra integral action and helps to reduce the control error:

$$\Delta u^{(j)} = k_{ss} \sum_{j=1}^{N_{pp}} e_s^{(j)}. \tag{4}$$

The tuning knob of this metric is k_{ss}, its value is user-defined. The follow-up models are then constructed iteratively, on a window of past N_{pp} point pairs $(\Delta u^{(j)}, e_s^{(j)})$ as new plant data are acquired. The number of points in the regression set has a large impact on the capability of the model to adequately describe data. Thus, having in mind that plant data are acquired with each sampling period, selecting a great value for N_{pp} can cause the model to become insensitive to the change in data. Contrary to this, constructing a model based on a small number of points can seriously harm its efficiency to capture the data's curvature.

The overall process of $\Delta u^{(j)}$ calculation can be divided in the following phases: (i) initialization, (ii) acquiring data for regression, (iii) calculation of the incremental value. In the initialization phase the parameters of the DUPID controller and the plant are set; in the second phase, the points for model construction are acquired. An error point $e_s^{(j)} = SP^{(i)} - PV^{(i)}$, is acquired every time the simulation time steps counter i is divisible with the current value of DV; in the third phase, the approach taken for calculation of the incremental value depends on the number of acquired points. If this number is less than N_{pp}, then the next incremental value is calculated as given with Equation (4). Otherwise, if the number of points is equal or greater than N_{pp}, then the next incremental value is calculated from Equation (3). These steps are repeated until the maximum allowed simulation time steps i_{max} are met. The complete process of incremental value computation is modelled by the flow chart diagram shown in Figure 3.

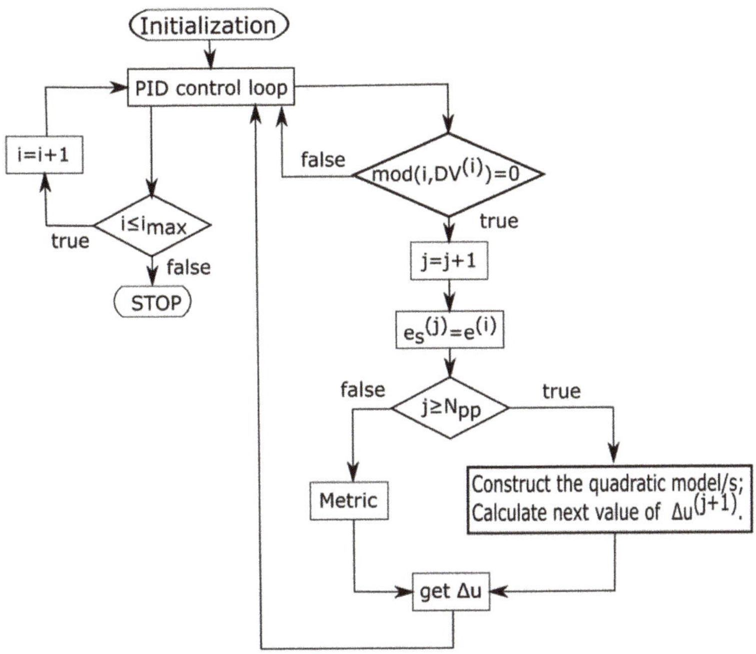

Figure 3. The flow chart diagram of the DUPID controller.

2.2.1. The 1D DUPID Controller

To ensure the quality of the regression data, this approach partitions the set $\mathbb{U}$ composed of N_{pp} points pairs $(\Delta u^{(j)}, e_s^{(j)})$ into two subsets with respect to the last gathered point $(\Delta u^{(k)}, e_s^{(k)})$ (red point fitted within triangle). The effect of data segmentation carried out by the 1D DUPID is shown in Figure 4.

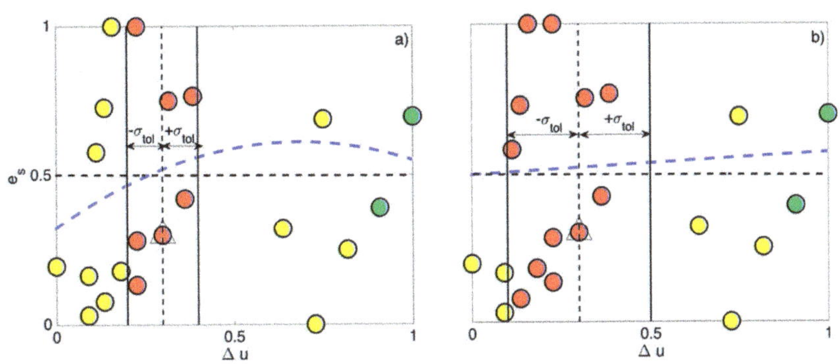

Figure 4. The effects of using of 1D DUPID algorithm to select adequate data for model construction for different values of σ_{tol}: **(a)** $\sigma_{tol} = 0.1$ and **(b)** $\sigma_{tol} = 0.2$. The blue dashed lines in **(a,b)** are the obtained quadratic fits. The last point fed to the algorithm is marked with a triangle. The collected data used for testing consist of 20 random integer points drawn on the interval $[0, 50]$ for Δu and $[-70, 70]$ for e_s. The data are normalized before it is fed to the algorithm. The black dashed line is zero error axis mapped in the normalized search space.

The last point obtained in the regression set will be addressed as the center point. The first subset, i.e., the vicinity set U_v, is comprised of all the points encompassed in $\pm\sigma_{tol}$ distance of the center point, measured on the Δu axis (red points). The point selection in this version of the algorithm is carried out solely with respect to the coordinate Δu (hence its "one-dimensional" nature). The second subset, the outer set U_o, contains the remaining points that are further from the center point $\Delta u^{(k)} \pm \sigma_{tol}$ (yellow and green points). The approach of segmenting the regression data is carried out to ensure the constructed model compromises between local precision on one hand and robustness on the other hand. The local precision of the model is guaranteed by the points found in the vicinity set U_v whereas the robustness is established by the points found in the outer set U_o. In this paper, the robustness of the fit is guaranteed by considering two anchor points (the green points) from the outer set. The anchor points are regarded as points whose distance with respect to $\Delta u^{(k)}$ is greatest. After the regression set $U_r = \{U_v, U_o\}$ is constructed, the model parameters $p = \{A, B, C\}$ are calculated from the optimization problem defined as:

$$\min_{p} \sum_{r=1}^{n_r} \left(e_s^{(r)} - E_M\left(\Delta u^{(r)}, p\right) \right)^2, \tag{5}$$

where, n_r is the number of points in the regression set, $\Delta u^{(r)}$ are the points found in U_r, the set of the past control error values is defined as $e_s = \left\{ e_s^{(1)}, \ldots, e_s^{(n_r)} \right\}$, and $E_M\left(\Delta u^{(r)}, p\right)$ describes the quadratic model given with Equation (3). Next, $\Delta u^{(j+1)}$ is calculated as:

$$\Delta u^{(j+1)} = \underset{\Delta u_i}{\operatorname{argmin}} \left| \Delta u_i - \Delta u^{(k)} \right|, \tag{6}$$

where Δu_i represent, the roots of Equation (3):

$$\Delta u_i = \{\Delta u_i |\ E_M = 0\}. \tag{7}$$

However, there might be a situation where the constructed fit does not intersect with the zero error axis. In such occasions $\Delta u^{(j+1)}$ is calculated by differentiation of the constructed model:

$$\Delta u^{(j+1)} = -\frac{B}{2A}. \tag{8}$$

Afterwards, $\Delta u^{(j+1)}$ is applied to the plant, and as a new error point $e_s^{(j+1)}$ is obtained, the array of past N_{pp} points is moved with center at $(\Delta u^{(j+1)}, e_s^{(j+1)})$ and the procedure for calculation of the next incremental value is repeated.

2.2.2. The 2D DUPID Controller

The process of selection of regression data in 2D DUPID is given in Figure 5. The 2D DUPID uses two coordinates, Δu and e_s, in the process of points selection. This is carried out by calculating the Euclidian distances of each of the $N_{pp} - 1$ points with respect to the center point $\left(\Delta u^{(k)}, e_s^{(k)}\right)$. As a result, a local coordinate system around the last point is placed (see Figure 5a):

$$D_m = \sqrt{\left(\Delta u^{(m)} - \Delta u^{(k)}\right)^2 + \left(e_s^{(m)} - e_s^{(k)}\right)^2}; m = 1, \ldots, N_{pp} - 1. \tag{9}$$

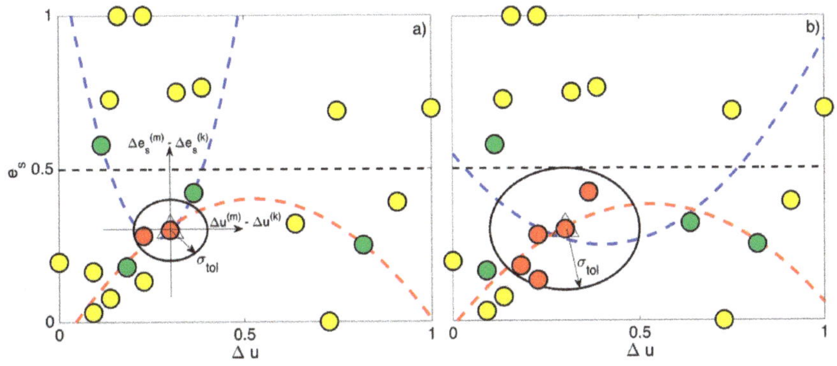

Figure 5. The effects of using of 2D DUPID algorithm to select adequate data for model construction for different values of σ_{tol}: (**a**) $\sigma_{tol} = 0.1$ and (**b**) $\sigma_{tol} = 0.2$. The blue and red dashed lines in (**a**,**b**) are the obtained quadratic fits. The last point fed to the algorithm is marked with a triangle. The collected data used for testing consist of 20 random integer points drawn on the interval [0, 50] for Δu and $[-70, 70]$ for e_s. The data are normalized before it is fed to the algorithm. The black dashed line is zero error axis mapped in the normalized search space.

At first, the vicinity set U_v consists of all the points encompassed by the circle centered at $\left(\Delta u^{(k)}, e_s^{(k)}\right)$ with radius σ_{tol} (red points):

$$\left(\Delta u^{(m)} - \Delta u^{(k)}\right)^2 + \left(e_s^{(m)} - e_s^{(k)}\right)^2 \leq (\sigma_{tol})^2. \quad (10)$$

The outer set U_o includes the points with distance greater than σ_{tol}. (green and yellow points). Those points are additionally sorted based on the angles they form with the center point. If the number of points present in the outer set is larger than four, then the algorithm proceeds with the segmentation of the outer set until a point is found in each local quadrant with the smallest distance D_m (green points). If the number of points is less than four, then all points are considered for regression. Finally, if the number of points found is exactly four, then two outer subsets are formed, each comprised of two points. The first outer subset U_o^1 contains the points found in the local quadrants 1 and 2, and the second outer subset U_o^2 contains the points found in the local quadrants 3 and 4. The major downside of this approach is that it may suffer in terms of time and computer resources needed for those four points to be found. The underlying idea of splitting the outer set into two subsets is to ensure that the data are fitted by two quadratic models (red and blue dashed lines), where at least one of them intersects with the zero error axis (black dashed line).

The subsets constructed from the outer set are defined as: $U_o^1 = \left\{\Delta u_o^{(1)}, \Delta u_o^{(2)}\right\}$ and $U_o^2 = \left\{\Delta u_o^{(3)}, \Delta u_o^{(4)}\right\}$. The values contained in both regression sets $U_r^1 = \{U_v, U_o^1\}$ and $U_r^2 = \{U_v, U_o^2\}$ are related to the entries found in the error sets, $e_s^{(1)(r)}$ and $e_s^{(2)(r)}$, respectively. The model parameters, $p^{(1)} = \{A_1, B_1, C_1\}$ and $p^{(2)} = \{A_2, B_2, C_2\}$ for both models are calculated from two optimizations problems, each defined as:

$$\min_{p^{(i)}} \sum_{r=1}^{n_r} \left(e_s^{(i)(r)} - E_M^{(i)}\left(\Delta u^{(i)(r)}, p^{(i)}\right)\right)^2, \; i = 1,2; \quad (11)$$

where n_r is the number of points in the regression set, $\Delta u^{(i)(r)}$ are the points found in $U_r^{(i)}$, $E_M^1\left(\Delta u^{(1)(r)}, p^{(1)}\right)$ is the quadratic model constructed from the $\left(U_r^1, e_s^{(1)(r)}\right)$ set, and $E_M^2\left(\Delta u^{(2)(r)}, p^{(2)}\right)$ is the model constructed based on the $\left(U_r^2, e_s^{(2)(r)}\right)$ set. Next, $\Delta u^{(j+1)}$ is calculated in a similar fashion as given with Equation (6), but the number of roots can vary depending on, whether the constructed models intersect with the zero error value axis or not. As a result, there might be three cases:

- Both constructed models intersect with the zero error axis. The total number of roots, in this case, is $i = 4$;
- Just one of the constructed models intersects with the zero error axis. The total number of roots, in this case, is $i = 2$;
- None of the constructed models intersects with the zero error value. Consequently, the next incremental value $\Delta u^{(j+1)}$ is chosen as the minimum value of the model that predicts the smallest error.

After $\Delta u^{(j+1)}$ is calculated, it is applied to the plant and a new error point $e_s^{(j+1)}$ is attained. The procedure of calculating the next incremental value is repeated by moving the coordinate system at $(\Delta u^{(j+1)}, e_s^{(j+1)})$.

3. Results

3.1. Case Study

In this paper, as a benchmark plant model for the controllers the highly nonlinear continuous stirred tank reactor (CSTR) system, Figure 6, with varying parameters is considered.

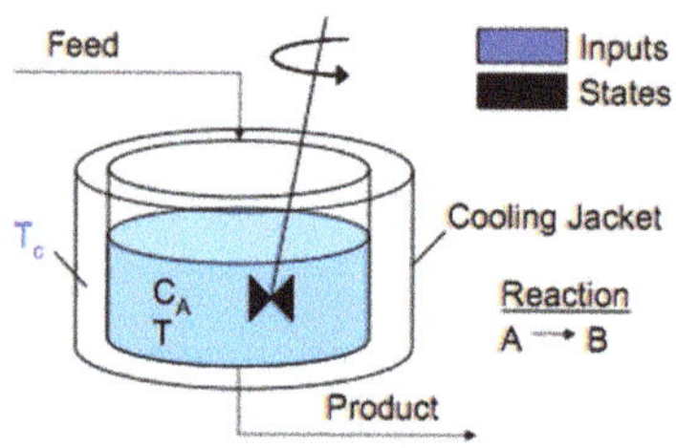

Figure 6. Graphical representation of the Continuous Stirred Tank Reactor (CSTR) system.

The model equations were taken from [20]. The CSTR system is a second-order control system which is modelled by the Equations (12) and (13). In spite of the simplicity of this model, it is well known by its strong nonlinear dynamics which is often manifested by: strong dependence of particular parameters, steady-state multiplicity, limit cycle etc. [23].

$$\frac{dT}{dt} = \frac{q}{V}\left(T_f - T\right) + \frac{\Delta H}{\rho C_p} k_0 C_A e^{-\frac{E}{RT}} + \frac{h A_{ta}}{V \rho C_p}(T_c - T), \tag{12}$$

$$\frac{dC_A}{dt} = \frac{q}{V}\left(C_{Af} - C_A\right) - k_0 C_A e^{-\frac{E}{RT}}. \tag{13}$$

The parameters of the model are defined as: q—Volumetric flow-rate; V—Volume of the CSTR; ρ—Density of the mixture; C_p—Heat capacity; ΔH—Heat of the reaction; E—Activation energy; R—Universal gas constant; k_0—Pre-exponential factor; h—Overall heat transfer coefficient; A_{ta}—Heat transfer area; T_f—Feed fluid temperature; C_{Af}—Feed concentration of reactant A; T_c—Temperature of the cooling jacket; C_A—Concentration of reactant A; T—Temperature in the CSTR. The nominal values assumed for the plant parameters are given in Table 1. The manipulated variable is assumed to be the cooling jacket fluid temperature T_c, which is operated in the interval [250, 350] K, and the reactor temperature T is assumed to be the controlled variable, and the only measured state.

Table 1. The considered nominal values for the parameters of the CSTR plant.

Parameter	Value	Parameter	Value
q	100 L/min	hA_{ta}	$5 \cdot 10^4$ J/min K
V	100 L	T_f	350 K
ρ	1000 g/L	C_{Af}	0.5 mol/L
C_p	0.239 J/g K	$T_c(0)$	300 K
ΔH	$5 \cdot 10^4$ J/mol	$C_A(0)$	≈0.46 mol/L
E/R	8750 K	$T(0)$	≈318.9 K
k_0	$7.2 \cdot 10^{10}$ 1/min	$[T_{c,\,min},\ T_{c,\,max}]$	[250, 350] K

For relevant comparison between the discussed control algorithms, it is assumed that the heat transfer coefficient h and the feed fluid temperature T_f change over time. These changes are realistic in practice as they often occur as a result of material deposition (fouling film) on the heat transfer surfaces, over the period of the plant operation. The process of fouling film development on the heat transfer surfaces is a slow and gradual process, which as a consequence adds resistance to the flow of heat. There are two types of fouling film build-up: asymptotic and linear. In the former, the resistance due to the fouling film increases very quickly at the beginning of the operation and becomes asymptotic to a steady-state value at the end. In the latter, the fouling film develops linearly over the entire period of plant operation [24].

Here, linear deposit contamination was assumed, and the parameters h and T_f were assumed to be subjected to linear change. The heat transfer coefficient h is replaced in Equation (12) by h_d, which is defined as:

$$h_d = \phi_h(t)h = (1 - \alpha_h t)h, \tag{14}$$

where t is time, h is the value of the heat transfer coefficient, h_d is the scaled version of the heat transfer coefficient, $\phi_h(t) \in (0, 1)$ is the fouling coefficient (the scaling coefficient), and α_h is the fouling constant. The feed fluid temperature T_f was assumed to be constant for a certain period of time, before changing into a ramp function in a negative direction. To broaden the discussion further, it was assumed that T_f changes as ramp function in positive direction also:

$$T_f = \begin{cases} T_{fconst};1 \leq i < lb \\ \pm T_d \frac{i-lb}{i_{max}-lb} + T_{fconst};\ lb \leq i \leq i_{max} \end{cases}, \tag{15}$$

where T_{fconst} is the initial value of the feed temperature, lb defines the interval of simulation time steps i while T_f is kept constant, T_d defines how much T_f changes in the positive or negative direction, and i_{max} is the maximal number of simulation time steps. The model equations are numerically solved in MATLAB by the function ode23t with simulation time step $T_i = 0.1$ min.

3.2. Simulation Results and Discussion

In this section, the control performances of the 1D and 2D DUPID controllers are compared against a benchmark PID controller. The benchmark PID controller is tuned by trial and error, as it is carried out in numerous industrial plants. To support this fact, in [25] the authors are referring to a statistic indicating that 30% of the PID controllers in the industry operate in manual mode and require continuous fine-tuning and supervising by the process technologist. Additionally, in [26], it is reported that 25% of the PID applications use coefficients that are pre-set by the manufacturer with no update of their values with respect to the particular process. Therefore, to showcase a realistic scenario, the parameters of the PID controller are tuned by trial and error assuming non-varying plant parameters. The PID controller parameters were assumed to be constant over the plant operation and they were selected so that the plant output achieves smooth transient response and zero steady-state error, without breaching the bounds of the interval of admissible control values. As a quantifying measure for the performance assessment of the controllers, the Integral of Absolute Error (IAE) metric is used (Equation (16)).

$$IAE = \frac{1}{i_{max}} \sum_{i=1}^{i_{max}} \left| e^{(i)} \right|. \tag{16}$$

For the sake of relevant comparison, three scenarios were addressed when the controllers were applied to control the CSTR system. These three scenarios are defined as follows:

- In the first scenario (Scenario 1), it is assumed that the heat transfer coefficient linearly drops to 50% of its initial value. Therefore, the value of α_h was assumed to be 0.0167.
- In the second scenario (Scenario 2), it is assumed that the feed temperature T_f linearly increases for $T_d = +40\ K$ of its initial value, $T_{fconst} = 350\ K$. The value of lb is 30.
- In the third scenario (Scenario 3), it is assumed that the feed temperature T_f linearly drops for $T_d = -40\ K$ of its initial value $T_{fconst} = 350\ K$. The value of lb is 30.

The PID parameters were tuned to be $K_p = 4.5$, $K_i = 3.31$ and $K_d = 0.01$. All DUPID parameters are pre-set and are constant in all scenarios. The DUPID parameters in both versions are assumed to be constant in all scenarios, $N_{pp} = 12$ and $\sigma_{tol} = 0.1$. The simplest guideline for tuning these two parameters would be to select the value of the first parameter (N_{pp}) to be at least three times larger (we used here four times larger number) than the minimum needed points for establishing the model given with Equation (3), while selecting the value of 0.1 for the latter parameter (σ_{tol}) will be suitable in most of the cases. Regarding the rest of the DUPID parameters, the k_{ss} value is set to be 1 and DV is given as:

$$DV = \begin{cases} DV^{(i)} = 10; i \in [1, 10] \\ DV^{(i)} = 1;\ i \in [11, i_{max}] \end{cases}. \tag{17}$$

The SM calculates the first incremental value after ten simulation time steps (one minute), whereas the other incremental values are calculated in each simulation time step. The idea behind the calculation of the first incremental value after ten simulation time steps is to start adding bias after the plant is settled in the steady-state. Adding bias during the transient state can provoke an unwanted overshoot in the plant response, or in the extreme case, it can destabilize the plant. The possible downside of calculating the other incremental values in each simulation time steps is having the added computational time burden. On the other hand, this can be justified by achieving a smaller IAE value and a better overall control performance.

In a real case scenario, the number of time steps needed to pass for the SM to start calculating the incremental value will be determined based on the plant itself, and on the experience of the process technician (the operator) on how much time is needed for the plant output to settle in steady-state. However, to omit the possible subjective human decision, in future modifications of the DUPID algorithm the moment of the first incremental value calculation should be determined automatically by accounting for the proximity of the PV with respect to the SP over time.

The simulation time step is assumed to be $T_i = 0.1$ min. In the first ten simulation time steps (until minute 1) the value of Δu is kept on zero since the first calculated incremental value by SM is carried out at the tenth time step. The termination criterion was defined with respect to the maximum permitted time steps $i_{max} = 300$. The total time of the plant operation is $i_{max} \cdot T_i = 30$ min.

In Figure 7, the resulting plant response under PID control in all three scenarios is given. It is indicative that the error area increases over time in all scenarios, (d), (e) and (f). This is a consequence of the changing plant parameters over time. Since the PID controller coefficients are kept constant the controller is unable to adapt to the change in plant operating conditions and is therefore unable to steer the controlled variable (T) to the set point.

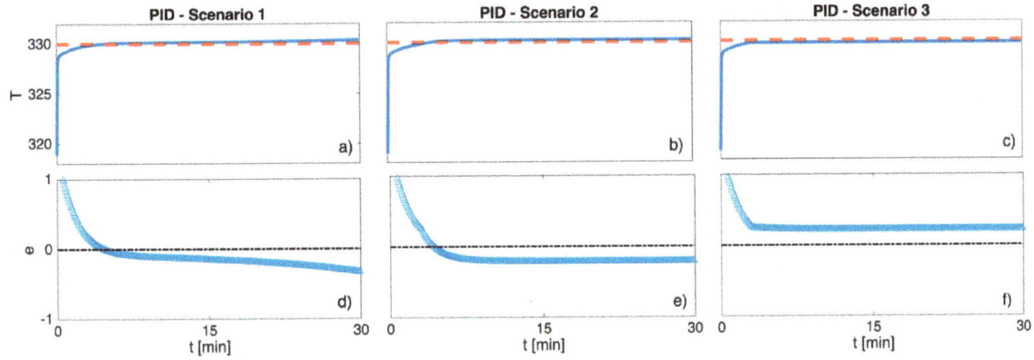

Figure 7. The plant response under PID control in all three scenarios. The figure is structured in two rows of three pictures, in the first row (**a–c**) show the plant output (blue line) and the set point (red dashed line); in the second row (**d–f**) show the resultant control error (blue markers) is plotted against the zero error line (black dashed line).

The 1D and 2D DUPID controllers proved to be more agile in detecting and countering the effects caused by the time-varying parameters. From the Figures 8d–f and 9d–f, it can be seen that the DUPID anticipates the change in plant parameters and pushes the PID control value in the right direction to compensate for its stationary coefficients.

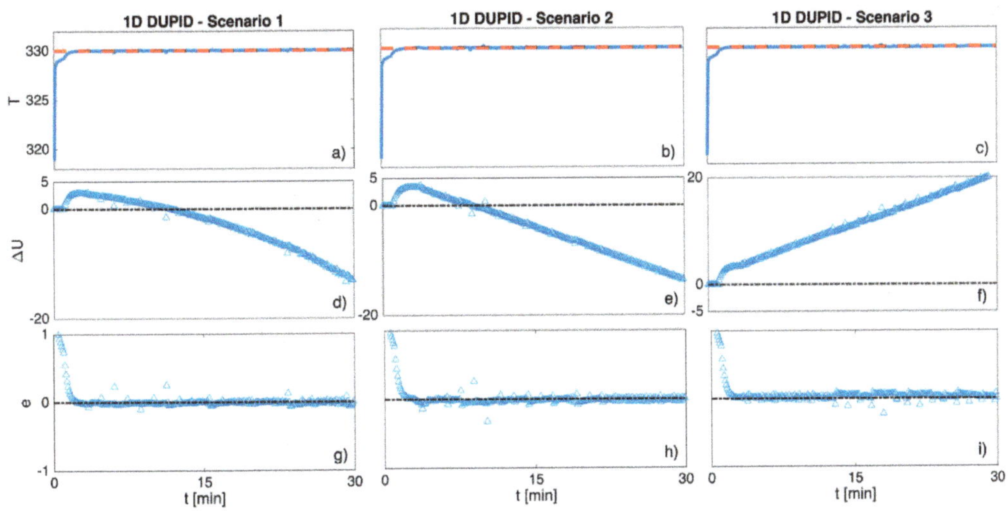

Figure 8. The plant response under 1D DUPID control in all three scenarios. The figure is structured in three rows of three pictures, in the first row (**a–c**) the plant output (blue line) and the set point (red dashed line) are plotted; in the second row (**d–f**) the incremental value is given (blue markers); in the third row (**g–i**) the resultant control error (blue markers) is plotted against the zero error line (black dashed line).

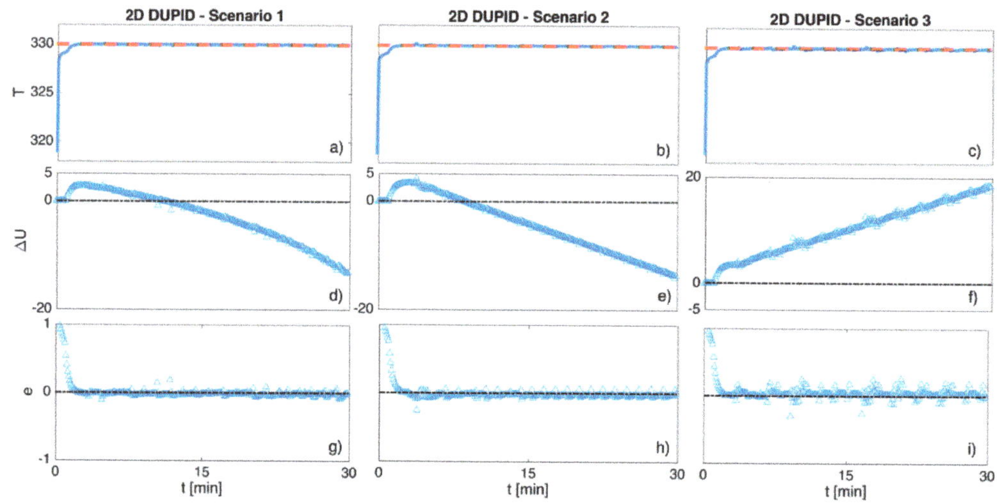

Figure 9. The plant response under 2D DUPID control in all three scenarios. The figure is structured in three rows of three pictures; in the first row (**a–c**) the plant output (blue line) and the set point (red dashed line); in the second row (**d–f**) the incremental value is given (blue markers); in the third row (**g–i**) the resultant control error is given (blue markers) and it is plotted against the zero error line (black dashed line).

The control performance of each controller was quantified by the IAE metric, Table 2b. The relative differences (The relative difference is calculated as $d_r = |x - y| / \max(|x|, |y|)$) of the obtained IAE values, Table 2c, were calculated for each scenario. On average, the 1D DUPID and 2D DUPID controllers achieved improvement over PID control performance of around 62% and 59%, respectively. The direct comparison between 1D and 2D DUPID controllers showed that 1D DUPID is performing slightly better, on average of around 7% (Table 2c) first column). Moreover, we compared 1D and 2D DUPID in terms of how much time is needed on average for each controller to calculate the incremental value in each scenario, Table 1. From the values of the calculated relative differences, it is clear that 2D DUPID needs around twice as much time on average to produce the next incremental value compared to the 1D DUPID. This is an expected result having in mind that the 2D DUPID has an added computational burden which comes from its design to search for a point in each of the four quadrants of the local coordinate system placed around the centre point. Overall, the two controllers 1D and 2D DUPID showed that they can steer the controlled variable toward the set point with satisfactory precision.

Table 2. Table (**a**) contains the average time needed for calculating the incremental value by each DUPID controller in the considered scenarios (Scen.). In the last column of this table, the relative differences (RD) are given. In table (**b**), the integral of absolute error (IAE) values obtained by the controllers in each scenario are given, and in the last table, (**c**), the relative differences of the obtained IAE values by the controllers are given.

(a)	**Time in sec.**		
Scen.	**1D**	**2D**	**RD**
1	0.0436	0.1007	0.5670
2	0.0497	0.1105	0.5502
3	0.0514	0.1070	0.5196

Table 2. *Cont.*

(b)		IAE values	
Scen.	1D	2D	PID
1	0.1002	0.1005	0.2529
2	0.1066	0.1084	0.2781
3	0.1097	0.1337	0.2960
(c)		RD values	
Scen.	1D/2D	1D/PID	2D/PID
1	0.0030	0.6038	0.6026
2	0.0166	0.6167	0.6102
3	0.1795	0.6294	0.5483
Avg.	0.0664	0.6166	0.5870

4. Conclusions

In this paper, we discussed an improved version of the PID controller, the Dynamically Updated PID controller. The DUPID controller is constructed based on the assumption that the parameters in the plant vary with time in a smooth and gradual fashion. Its objective is to track the variation of the parameters and to intervene in the PID control value to reduce the increasing control error, which is a direct consequence of their variation. This controller resides on the paradigm of using recent history error data to construct a quadratic model that models the error caused by the varying parameters. This model is then exploited for the iterative calculation of a sequence of incremental values, which are applied directly to the PID control value, without changing the controller parameters, for the purpose of lessening the effect caused by the varying parameters in plant's output.

The control performance of the DUPID controller is dependent on the precision of the quadratic model. To ensure the quadratic model describes the data well, the regression data used for its construction must fulfil certain criteria. In this regard, we discussed two versions of DUPID, the 1D and 2D DUPID. These versions differ in the approach of how the data are selected before the model coefficients are determined: the 1D DUPID uses only one coordinate in the process of regression data selection, whereas the 2D DUPID uses two coordinates for data selection.

We compared the control performances of the conventional parallel-form PID controller and the DUPID controllers (1D and 2D DUPID). Both controllers were applied to the CSTR system with varying parameters. We observed the control performance of the controllers in three scenarios which were derived from the assumption that the parameters h and T_f vary linearly over the time the plant is controlled. The results showed that the DUPID controller was capable of reducing the control error to a satisfactory level, whereas the PID control performance dropped with time. Finally, we performed a direct comparison between 1D and 2D DUPID controllers, and the results showed that their control performances do not differ by much.

To illustrate that the DUPID controller has the potential of broader use, different than the one discussed here, a meaningful control scenario is pinpointed in the Supplementary Materials. Namely, the DUPID controller is used to attribute the problem of the poorly tuned integral term in the PID controller. In this scenario, it was assumed that the integral term (K_i) of the PID controller is seriously mistuned and the set-point is changing as a ramp function. The results indicated that the DUPID controller accomplished set-point tracking with satisfactory steady-state error, and therefore it was able to make up for the mistuned integral term. To show that the improvement is genuine, we simulated the plant for 10 different uniformly distributed values of K_i drawn from the interval $[0.01, 1]$.

For all the realizations of K_i the DUPID showed superior performance with respect to the mistuned PID controller.

There are few open questions that will be addressed in the future versions of the DUPID controller:

First of all, to prove that the DUPID controller is multifaceted, it should be tested on the same plant discussed here, but when the concentration of the reactant A (C_{Af}) or the volumetric flow rate (q) are subject to change. Moreover, it will be tested on more complex high-order nonlinear processes of practical importance. In the next modifications, the DUPID algorithm will be adapted to be able also to detect and tackle step changes in the plant parameters.

In addition, a trust-region based on covariance analysis around the centre point will be included. The importance of the trust-region will be twofold; it will constrain the algorithm to calculate an incremental value in the direction of poor level data spread and, at the same time, it will define bounds of the validity of the model.

Lastly, after the aforementioned improvements are implemented, the DUPID controller will be tested based on its ability to average out the noise that is omnipresent in real plants.

Furthermore, one should note that in this paper, the focal point was set on discussing the hands-on implementation of the DUPID controller without elaborating on the stability margins of the proposed algorithm. However, the topic of stability is a significant issue in control engineering, so further investigations will be conducted on establishing a formal mathematical proof of the asymptotical stability of the plant around a given set-point when DUPID is used as controller. This discussion is motivated by the result reported in [27], where the authors proved that for CSTRs that are isothermal and asymptotically stable, a classical PI compensator can yield to global asymptotic stabilization about a prescribed operating point [23].

Supplementary Materials: A broader elaboration on the control scenario discussed in Section 4 (Conclusions), paragraph 4, is available online at https://www.mdpi.com/1999-4893/14/2/31/s1.

Author Contributions: Conceptualization, D.S.; Investigation, D.S.; Methodology, D.S; Software, D.S.; Supervision, G.N., S.D., M.S.; Validation, G.N., S.D., M.S.; Writing-original draft, D.S.; Writing-review and editing, D.S. and G.N. All authors have read and agreed to the published version of the manuscript.

Funding: This research received no external funding.

Conflicts of Interest: The authors declare no conflict of interest.

References

1. Åström, K.J.; Hägglund, T.; Astrom, K.J. *Advanced PID Control. vol. 461. ISA-The Instrumentation, Systems, and Automation Society Research Triangle.* 2006. Available online: https://www.researchgate.net/publication/3207690_Advanced_PID_Control_-_Book_Review (accessed on 19 January 2021).
2. Rotach, V.Y. *Teoriya Avtomaticheskogo Upravleniya: Uchebnik dlya VUZov (Automated Control Theory: University Textbook)*, 5th ed.; MEI Publishing House: Moscow, Russia, 2008.
3. Fahmy, R.A.; Badr, R.I.; Rahman, F.A. Adaptive PID Controller Using RLS for SISO Stable and Unstable Systems. *Adv. Power Electron.* **2014**, *2014*, 1–5. [CrossRef]
4. Ziegler, J.G.; Nichols, N.B. Optimum Settings for Automatic Controllers. *J. Dyn. Syst. Meas. Control.* **1993**, *115*, 220–222. [CrossRef]
5. Haykin, S.S. *Adaptive Filter Theory*; Pearson Education India: Chennai, India, 2008.
6. Åström, K.J.; Wittenmark, B. *Adaptive Control*; Courier Corporation: North Chelmsford, MA, USA, 2013.
7. Tian, B.; Su, H.; Chu, J. An optimal self-tuning PID controller considering parameter estimation uncertainty. In Proceedings of the 3rd World Congress on Intelligent Control and Automation (Cat. No.00EX393), Hefei, China, 26 June–2 July 2000; pp. 3107–3111. [CrossRef]
8. Martínez, J.I.O.; Paz Ramos, M.; Garibo, S.; Luna-Ortega, C. PI controller with dynamic gains calculation to comply with time specs in presence of parametric disturbances. In Proceedings of the 6th International Conference Electrical Engineering, Computing Science and Automatic Control, Toluca, Mexico, 10–13 November 2009.
9. Xiao, S.; Li, Y.; Liu, J. A model reference adaptive PID control for electromagnetic actuated micro-positioning stage. In Proceedings of the 2012 IEEE International Conference on Automation Science and Engineering (CASE), Seoul, Korea, 20–24 August

2012; pp. 97–102. Available online: http://citeseerx.ist.psu.edu/viewdoc/download?doi=10.1.1.721.2583&rep=rep1&type=pdf (accessed on 19 January 2021).
10. Petersen, I.R.; Tempo, R. Robust control of uncertain systems: Classical results and recent developments. *Automatica* **2014**, *50*, 1315–1335. [CrossRef]
11. Vilanova, R.; Alfaro, V.M.; Arrieta, O. Robustness in PID Control. In *PID Control in the Third Millennium*; Springer: London, UK, 2012; pp. 113–145.
12. Morari, M.; Zafiriou, E. *Robust Process Control*. 1989. Available online: https://cse.sc.edu/~{}gatzke/cache/morari-zafiriou-PI-chapters1-3.pdf (accessed on 19 January 2021).
13. Lee, Y.; Park, S.; Lee, M.; Brosilow, C. PID controller tuning for desired closed-loop responses for SI/SO systems. *AIChE J.* **1998**, *44*, 106–115. [CrossRef]
14. Ge, M.; Chiu, M.-S.; Wang, Q.-G. Robust PID controller design via LMI approach. *J. Process. Control* **2002**, *12*, 3–13. [CrossRef]
15. Garpinger, O.; Hägglund, T. A Software Tool for Robust PID Design. *IFAC Proc. Vol.* **2008**, *41*, 6416–6421. [CrossRef]
16. Mercader, P.; Astrom, K.J.; Banos, A.; Hagglund, T. Robust PID Design Based on QFT and Convex–Concave Optimization. *IEEE Trans. Control. Syst. Technol.* **2016**, *25*, 441–452. [CrossRef]
17. Stavrov, D.; Nadzinski, G.; Stojanovski, G.; Deskovski, S. Performance analysis of DUPID control over CSTR sys-tem with varying parameters. In Proceedings of the 14th International Conference ETAI, Struga, Macedonia, 20–22 September 2018; pp. 2545–4889.
18. Wenzel, S.; Gao, W.; Engell, S. Handling Disturbances in Modifier Adaptation with Quadratic Approximation. The research leading to these results was funded by the ERC Ad-vanced Investigator Grant MOBOCON under the grant agreement No. 291458. *IFAC-PapersOnLine* **2015**, *48*, 132–137. [CrossRef]
19. Wenzel, S.; Paulen, R.; Engell, S. Quadratic Approximation in Price-Based Coordination of Constrained Systems-of-Systems. Focapo/cpc. 2017. Available online: http://folk.ntnu.no/skoge/prost/proceedings/focapo-cpc-2017/FOCAPO-CPC%202017%20Contributed%20Papers/92_CPC_Contributed.pdf (accessed on 19 January 2021).
20. Seborg, D.E.; Mellichamp, D.A.; Edgar, T.F.; Doyle, F.J., III. *Process Dynamics and Control*; John Wiley & Sons: Hoboken, NJ, USA, 2010.
21. O'Dwyer, A. *Handbook of PI and PID Controller Tuning Rules*; Imperial College Press: London, UK, 2009.
22. Brunton, S.L.; Kutz, J.N. *Data-Driven Science and Engineering: Machine Learning, Dynamical Systems, and Control*, 1st ed.; Cambridge University Press: Cambridge, UK, 2019.
23. Alvarez, J.; Alvarez-Ramirez, J. Linear PI control of batch exothermic reactors with temperature measurement. *Int. J. Robust Nonlinear Control* **2006**, *16*, 113–131. [CrossRef]
24. Nikravesh, M.; Farell, A.E.; Stanford, T.G. Control of nonisothermal CSTR with time varying parameters via dy-namic neural network control (DNNC). *Chem. Eng. J.* **2000**, *76*, 1–16. [CrossRef]
25. Bucz, Š.; Kozáková, A. Advanced methods of PID controller tuning for specified performance. *PID Control Ind. Process* **2018**, 73–119. Available online: https://www.intechopen.com/books/pid-control-for-industrial-processes/advanced-methods-of-pid-controller-tuning-for-specified-performance (accessed on 19 January 2021).
26. Yu, C.-C. *Autotuning of PID Controllers: A Relay Feedback Approach*; Springer Science & Business Media: Berlin, Germany, 2006.
27. Alvarez-Ramirez, J. Global stabilization of chemical reactors with classical PI control. *Int. J. Robust Nonlinear Control IFAC Affil. J.* **2001**, *11*, 735–747. [CrossRef]

algorithms

MDPI

Article

Swarm-Based Design of Proportional Integral and Derivative Controllers Using a Compromise Cost Function: An Arduino Temperature Laboratory Case Study

P. B. de Moura Oliveira [1,2,*], John D. Hedengren [3] and E. J. Solteiro Pires [1]

1 Department of Engineering, University of Trás-os-Montes and Alto Douro (UTAD), 5000-801 Vila Real, Portugal; epires@utad.pt
2 Institute for Systems and Computer Engineering, Technology and Science (INESC-TEC), Campus da Faculdade de Engenharia da Universidade do Porto (FEUP), 4200-465 Porto, Portugal
3 Department of Chemical Engineering, Brigham Young University, Provo, UT 84602, USA; john.hedengren@byu.edu
* Correspondence: oliveira@utad.pt

Received: 25 October 2020; Accepted: 27 November 2020; Published: 30 November 2020

Abstract: Simple and easy to use methods are of great practical demand in the design of Proportional, Integral, and Derivative (PID) controllers. Controller design criteria are to achieve a good set-point tracking and disturbance rejection with minimal actuator variation. Achieving satisfactory trade-offs between these performance criteria is not easily accomplished with classical tuning methods. A particle swarm optimization technique is proposed to design PID controllers. The design method minimizes a compromise cost function based on both the integral absolute error and control signal total variation criteria. The proposed technique is tested on an Arduino-based Temperature Control Laboratory (TCLab) and compared with the Grey Wolf Optimization algorithm. Both TCLab simulation and physical data show that satisfactory trade-offs between the performance and control effort are enabled with the proposed technique.

Keywords: PID control; particle swarm optimization; grey wolf optimization; arduino

1. Introduction

Despite the development of more refined control techniques, the Proportional, Integral, and Derivative (PID) control continues to be ubiquitous for industrial control [1,2]. Given the practical relevance of this type of control, many design methods have been proposed since the pioneering work developed by Ziegler and Nichols [3]. PID tuning rules were developed for specific aspects, such as: Control modes (P, PI, PD, or PID); types of system model characteristics or forms (first order plus time delay, second order plus time delay, non-minimum phase, oscillatory, etc.); PID controller structure (parallel, series, with filters, with two-degrees of freedom, etc.); and anti-windup schemes [4–7]. The PID control practical relevance also motivates inclusion in most introductory feedback control courses, for example [8].

Artificial neural networks, fuzzy logic, and evolutionary computation have been successfully applied for the PID controller design. The ever-increasing computational power enables a fast practical implementation of computer-based PID controller design methodologies. Optimization-based techniques have advantages over classical tuning methodologies, as the former can be used independently of both system dynamics and PID control structure. Examples of the most well-established population-based algorithms which seek inspiration from natural phenomena are: Genetic Algorithm (GA) [9], Ant-Colony Optimization, (ACO) [10], Genetic Programming (GP) [11], Differential Evolution

(DE) [12], Particle Swarm Optimization (PSO) [13], Cuckoo Search Algorithm (CS) [14], Firefly Algorithm (FA) [15], Glowworm Swarm Optimization (GSO) [16], Gravitational Search Algorithm (GSA) [17], Grey Wolf Optimization (GWO) [18–21], and Elephant Herding Optimization (EHO) [22]. From this set, the ones most used within control design applications are GA [23,24] and more recently the PSO [25,26]. PSO has been improved as a result of significant research effort [27,28].

A commonly used argument against using bio-inspired and nature inspired metaheuristics over classical PID design methods is that these also require the adjustment of parameters prior to the optimization procedure. An advantage of the classical PSO algorithm is fewer adjustable heuristic settings compared with the basic GA. Moreover, PSO is more straightforward to implement than a GA. As the proposed technique aims to be simple to configure, the PSO algorithm is the selected optimization tool used in this study. Moreover, the proposed technique control performance is compared with a more recently introduced metaheuristic: The Grey Wolf Optimization proposed by [18].

There is an increasing pedagogical and research benchmark interest in pocket-sized and portable-based control experiments, as revealed by the following examples in [29–34]. Microcontrollers enable a wide range of pocket-size laboratories as common tools both for teaching/learning purposes, as well as for control engineering practitioners testing controller designs. The paradigm is shifting from monolithic laboratory experiments that require significant resources to take-home and modular experiments. The Temperature Control Laboratory (TCLab) proposed by [35–38] is used to teach PID control to undergraduate engineering students [39,40]. The TCLab is a low-cost Arduino-based lab which is a pocket-sized, plug-and-play kit, meaning that it does not require the user to perform the assembly. This is quite advantageous for computer-based control courses in which the primary goal is to test the identification and control techniques rather than hardware assembly. The TCLab is programmed using different environments, such as: Python, MATLAB®/SIMULINK®, and GNU Octave. Moreover, a wide range of open-source supporting materials, including programs, videos, and tutorials are freely available in [24].

The PSO was applied to identify a first order plus time delay model for the TCLab kit in [39,40]. In this paper, the PSO design of PID controllers based on an additive compromise cost function involving the Integral Absolute Error (IAE) and Total Variation (TV) is proposed. This technique is validated both with simulation and practical results obtained with the TCLab. The results obtained with the proposed technique are compared with the ones obtained with the GWO algorithm and classical tuning techniques. The novel contributions of this paper are highlighted in the following points:

- New formulation for an additive compromise (or aggregated) cost function involving the Integral Absolute Error (IAE) and Total Variation (TV). Major control design criteria concern optimizing set-point tracking and disturbance rejection, while minimizing the control signal variation. This proposed technique significantly simplifies the PID controller design procedure combining these criteria into a single-objective optimization formulation.
- PSO algorithm to design PID controllers that minimize a cost function weighting IAE and TV. A simple PSO algorithm constitutes an excellent design tool for the PID controller, with practical interest for control engineers. The proposed technique is compared with the original GWO algorithm, in a TCLab temperature control case study, providing a similar control performance.
- Both the simulation and practical validation with TCLab tests, show an effectiveness to design PID controllers by softening the control signal.

The paper remainder is organized as follows: Section 2 introduces the problem statement. In Section 3, the classical particle swarm optimization algorithm is presented. Sections 4 and 5 discuss the TCLab main features and results. Finally, Section 6 provides the conclusions and recommendations for future work.

2. PID Control Design: Problem Statement

Consider a general block diagram representation of a closed-loop system in Figure 1, assuming the following representation where r is the reference input (set-point), y is the controlled output, e is the error signal, u is the control output, d_1 is a load disturbance, d_2 is an output disturbance, n is a noise signal, and G_c and G_p are the controller and process transfer functions.

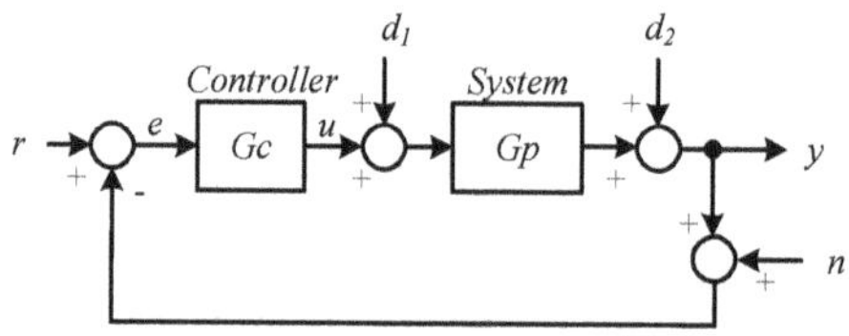

Figure 1. General closed-loop control system block diagram.

The ideal feedback control corresponds to obtaining the controlled output equal to the reference input independent of system disturbances. This ideal control is not possible in practice due to disturbances, measurement noise, actuator saturation, and a wide range of other physical limitations. With these limitations, the controlled output should attempt to follow the reference input (Set-Point Tracking, SPT), while rejecting the system disturbances (Disturbance Rejection, DR). In this work, it is assumed that a dynamic model of the system can be identified. In addition to the SPT and DR criteria, it is also important to consider the control effort in the controller design. A relevant research question is: How can a PID controller be designed to achieve the best possible set-point performance while minimizing the control effort?

There are many time-domain criteria to evaluate a controller performance. These criteria frequently involve set-point step responses with the smallest possible values for first overshoot and rise time. An indirect approach to simultaneously minimize several step response indices is using error-based criteria. The most common are the following: Integral of Square Error (ISE), Integral of Absolute Error (IAE), Integral of Time Weighted Absolute Error (ITAE), and Integral Weighted Square Error (ITSE), represented respectively by the following expressions:

$$ISE = \int_0^{t_{\text{sim}}} e^2(t)\, dt \tag{1}$$

$$IAE = \int_0^{t_{sim}} \left|e(t)\right| dt \tag{2}$$

$$ITAE = \int_0^{t_{sim}} t\left|e(t)\right| dt \tag{3}$$

$$ITSE = \int_0^{t_{sim}} te^2(t) dt \tag{4}$$

with t_{sim} representing the simulation time, which should be large enough so that the system-controlled output reaches the steady-state value. Depending on which of these criteria is adopted in the optimization cost function, different PID controller settings are obtained. Among these criteria the most widely used are the ISE, IAE, and ITAE. Time weighting the absolute error penalizes the error as time increases for ITAE, promoting the elimination of small steady-state errors. IAE independently gives the same relevance to errors over time. For this reason, IAE is adopted in this study. A common criterion to measure the control effort is:

$$E_u = \int_0^{t_{sim}} \left|u(t)\right| dt \tag{5}$$

with u representing the control signal. The control signal smoothness can be evaluated using a Total Variation (TV) index (6), approximated by (7):

$$TV_u = \int_0^{t_{sim}} \left|\frac{du}{dt}\right| dt \quad (6)$$

$$TV_u = \sum_{k=0}^{\frac{t_{sim}}{h}} |u(k+1) - u(k)| \quad (7)$$

with h representing the measurement sampling interval. The TV criterion is adopted in this study to minimize the control signal variation. It is worthwhile to mention at this point, that all actuators have physical limitations, and it is common that the control signal can vary linearly in the interval $[u_{min}, u_{max}]$, with u_{min} and u_{max} representing the minimum and maximum control signal values, respectively. A realistic controller design for practical implementation should consider specific actuator saturation limits. Measurement noise issues are not addressed in this study.

With two performance criteria selected (IAE and TV), a decision regarding how to use these in the optimization algorithm must be made. The two criteria can be considered as two separate functions within a multi-objective (or many-objective) optimization problem and solved using a PSO Pareto-based approach [41]. However, depending on the number of objectives, the computational burden associated with multicriteria optimization can be significant. After a non-dominated Pareto front is achieved, a decision support system should be used to help the operator decide the appropriate trade-offs from the Pareto front. Considering that just two criteria are used (IAE and TV), a simpler approach uses an additive compromise cost function. This cost function is represented by:

$$J = \alpha\, IAE + \beta\, TV \quad (8)$$

with α and β representing weighting factors determining the relative importance of each criterion. An equivalent single parameter β could also be used if the α value is always fixed to a value of one as done in this paper.

3. Classical Particle Swarm Optimization

The particle swarm optimization algorithm proposed by Kennedy and Eberhart [13] is inspired by the animal swarms social behavior, such as bird flocks and fish schools. The basic notion of the algorithm can be explained using the illustrative example presented in Figure 2.

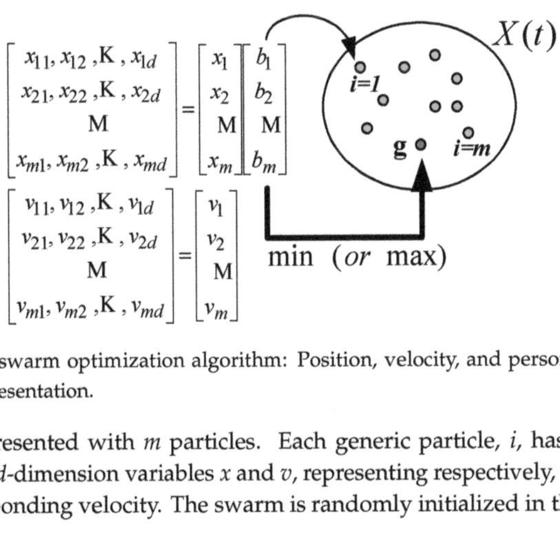

Figure 2. Particle swarm optimization algorithm: Position, velocity, and personal and global best variables data representation.

A swarm is represented with m particles. Each generic particle, i, has dynamics which are characterized by two d-dimension variables x and v, representing respectively, the particle position in space and the corresponding velocity. The swarm is randomly initialized in the search space unless there is specific information regarding the problem. The specific information can be incorporated in the initialization procedure to improve the convergence rate to an optimal solution. Each PSO iteration, t, updates every particle position to a new position. The update is from the sum of a velocity term according to (8) with the velocity evaluated using (9):

$$x_i(t+1) = x_i(t) + v_i(t+1) \tag{9}$$

$$v_i(t+1) = \omega\, v_i(t) + c_1\phi_1[b_i(t) - x_i(t)] + c_2\phi_2[g(t) - x_i(t)] \tag{10}$$

where b_i represents the best position obtained by particle i until the current iteration, t; g represents the best position from the full swarm (global best) or a specific neighborhood (local best). In this study, information is shared among the entire swarm (fully-connected model). As it can be observed from (9), the new velocity value is evaluated by a sum of three parts. The first part ($\omega\, v_i(t)$) is the inertia and considers the previous velocity value with a factor, ω, called inertia weight. The second part ($c_1\phi_1[b_i(t){-}x_i(t)]$) represents the particle cognitive knowledge update. It is the difference between the particle i individual best position, b_i, and the particle current position, x_i, multiplied by a cognitive constant c_1 and a disturbing random value, ϕ_1. The third part ($c_2\phi_2[g(t) - x_i(t)]$) represents the particle social knowledge update: The difference between the swarm global best position, g, and the particle current position, x_i, multiplied by a social constant c_2 and a disturbing random value, ϕ_2. The same relevance is usually given to the cognitive and social knowledge, and thus, c_1 and c_2 take the same value (e.g., 2 or 1.49445). The two random numbers, ϕ_1 and ϕ_2, are uniformly generated in the interval [0, 1]. Algorithm 1 is represented by the following pseudo-code:

Algorithm 1 PSO algorithm

```
1. t = 0
2. Initiate m,ω
3. Initialize swarm X(t)
4. Evaluate X(t)
while (not (termination criterion))
5. determine personal and global bests
for i = 1 to m
        6. Update v_i(t + 1)
        7. Update x_i(t + 1)
endfor
8. Update ω
9. t = t + 1
endwhile
```

As it can be observed from the PSO pseudo-code, this algorithm is conceptually quite simple. Regarding the PSO algorithm parametrization, there are three adjustable parameters: The swarm size, m, the termination criterion and the inertia weight limits. While these parameters are problem dependent, they can easily be selected for most PID control design cases. The swarm size should be large enough to guarantee a diverse solution representation across the search space. The termination criterion is often a pre-defined number of iterations and is adopted in this study. Regarding the inertia weight, it is commonly accepted that it should assume a higher value at the beginning of the search (e.g., $\omega_i = 0.7$) and then gradually decreased over the search evolution until it reaches a minimum value (e.g., $\omega_f = 0.4$). This inertia weight variation is to establish an important trade-off between space exploration in an early search stage and gradually transition to a global minimum consensus in a final search stage. The inertia weight is linearly decayed between the maximum and minimum values over the search iterations. However, other approaches can also be used (e.g., see [42]).

4. Temperature Control Laboratory (TCLab)

The Temperature Control Laboratory (TCLab) is an Arduino-based kit developed by [35]. It comes with the hardware set presented in Figure 3: An Arduino Leonardo; a Printed Circuit Board (PCB) with two transistors acting as heaters, two thermistors to sense the temperature in each transistor casing, an Light Emmiting Diode (LED) to indicate when the system is hot, and a connection socket to power the PCB components (see the white USB adapter shown in Figure 3); and finally, a Universal Serial Bus (USB) cable connects the Arduino to a computer (see the blue cable shown in Figure 3).

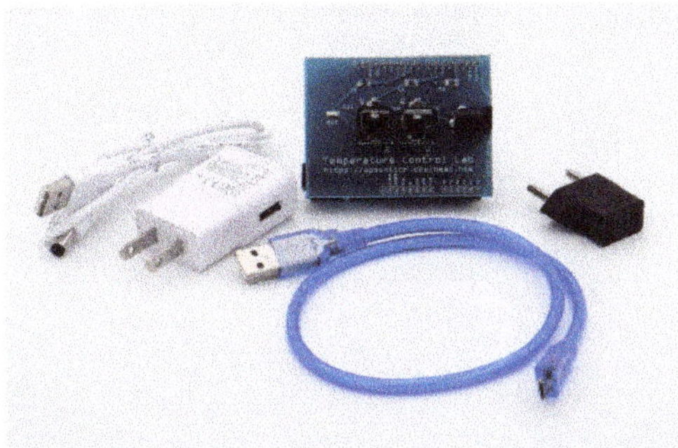

Figure 3. Temperature control lab (TCLab) hardware set.

The TCLab is presented from another perspective in Figure 4. As it can be observed, the transistors (TIP31C) are the control system actuators and operate as temperature heaters (Q1 and Q2). Two TMP36 sensors measure the temperature in each transistor casing (T1 and T2). As it can also be observed from Figure 4, the two transistors are connected to heat sinks acting as heat dissipators. The sensors are glued to the transistor casings using a non-conductive epoxy containing a thermochromic paint. This paint changes color when the temperature rises above 37 °C. The TCLab can be programmed using MATLAB®/SIMULINK®, Python and GNU Octave, with a wide range of programs freely available in [22]. This study uses the MATLAB®/SIMULINK® connection. More information regarding TCLab can also be found in [35–38].

Figure 4. TCLab with power and USB connections. Actuators and sensors outlined with letters A and S, respectively.

5. TCLab Results

Designing PID controllers with the proposed PSO technique involves (a) identifying a First-Order Plus Time Delay (FOPTD) model using an open-loop step response, (b) based on the obtained model in (a), another PSO algorithm is used to tune the PID controller. These two steps are illustrated in Figure 5 with the TCLab system.

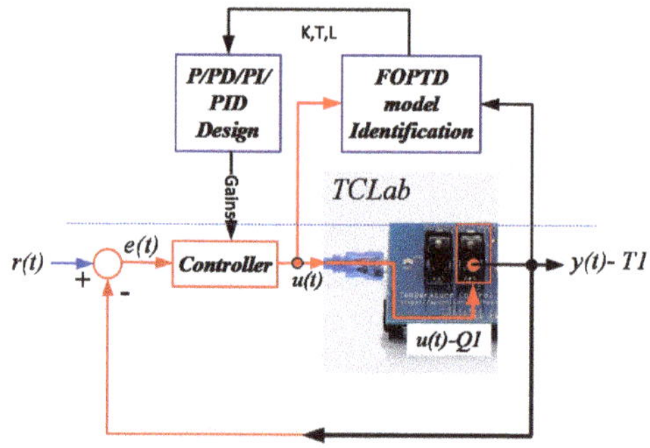

Figure 5. TCLab system identification and controller tuning approach.

5.1. *First-Order Plus Time Delay Model Identification*

The first issue that must be addressed is to identify a dynamic model relating temperature measured in transistor 1 (T_1) with the actuating heater signal (Q_1). A FOPTD model was identified based on an open-loop step response using both the classical two-point and PSO optimization methods as proposed in [39,40]. The following PSO conditions were used in the model identification tests:

- A swarm with size $m = 40$ particles.
- Each simulation was run for 50 iterations.
- The search intervals used for the controllers gains K, T, and L are: [0.1 3], [20 s 160 s], and [4 s 45 s], respectively. The FOPTD model parameters obtained with a classical step-response method (e.g., two-point method (see [28])) can be used to define the search interval.
- The inertia weight was initial and final values were $\omega_{\text{init}} = 0.7$ and $\omega_{\text{fin}} = 0.4$. These values were deemed appropriate with 50 iterations.

Figure 6 is a step response of the TCLab open-loop response. This figure top plot is a step applied to the system input (heater Q_1) with amplitude ranging from 0–80% applied at instant $t = 10$ s. The bottom plot is the result obtained with a FOPTD model identified using a PSO. The model [30,40] is:

$$G_p(s) = \frac{0.78}{1 + 152\,s} e^{-19.7s} \tag{11}$$

5.2. *PID Controller Tuning*

This section presents both simulation and TCLab results regarding the PID controller gains optimized with the PSO algorithm as depicted in the previous sections. The results are compared with the Cohen-Coon (CC) settings [43], as previously introduced in [39,40] and a more recent technique: AMIGO [5], which is well-known to achieve system robustness. For the sake of experiment replication, these two tuning rule methods for PID controllers are presented in Table 1.

Based on the FOPTD PSO model, PID controllers can be designed using several tuning methods. In [40], results were presented using the Cohen-Coon (CC) settings as these perform well for the TCLab. The results obtained with the CC PID settings $Kp = 13.47$, $Ti = 45.99$ s, and $Td = 7.00$ s are presented in Figure 7. This figure presents the overlapped results between the simulated and TCLab responses. A step input is applied to the system reference input at $t = 0$ s to define the TCLab transistor 1 set-point temperature at 25 °C and another one at t = 100 s increasing the set-point to 60 °C. A load disturbance with a magnitude of −40% is applied to the controller output at t = 380 s. Both the FOPTD model and the CC PID gains are used in the next section for comparison purposes with the other methods.

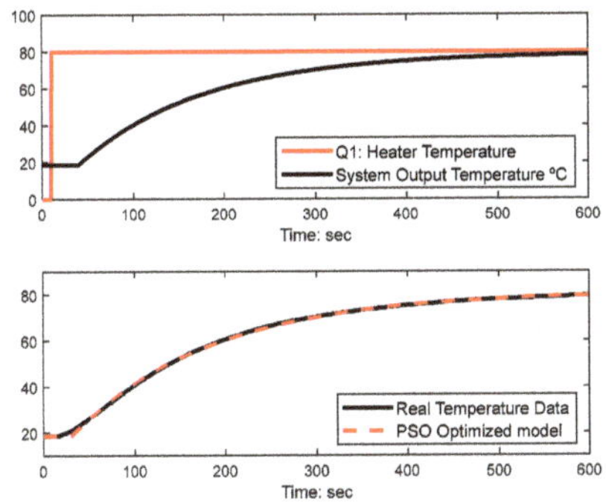

Figure 6. TCLab open-loop step response (**top plot**) and particle swarm optimization (PSO) model step response overlapped with the real TCLab data (**bottom plot**).

Table 1. Cohen-Coon [32] and AMIGO [5] proportional, integral, and derivative (PID) tuning rules.

Method	*Kp*	*Ti*	*Td*
Cohen-Coon	$\frac{1}{K}\frac{T}{L}\left(\frac{4}{3}+\frac{L}{4T}\right)$	$L\frac{32+6\frac{L}{T}}{13+8\frac{L}{T}}$	$L\frac{4}{11+2\frac{L}{T}}$
AMIGO	$\left(\frac{1}{K}\right)\left(0.2+0.45\frac{T}{L}\right)$	$\left(\frac{0.4L+0.8T}{L+0.1T}\right)L$	$\frac{0.5LT}{0.3L+T}$

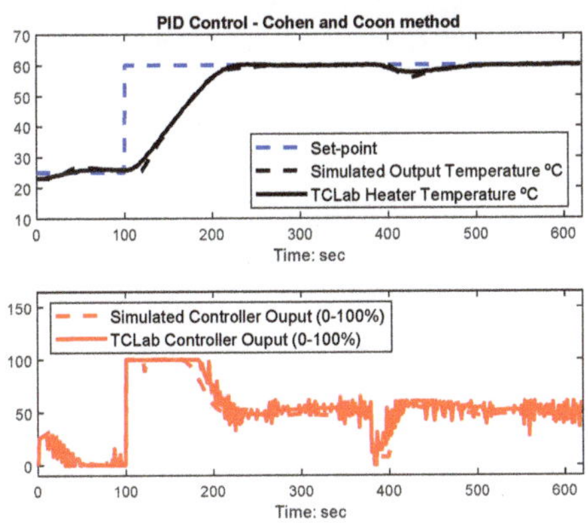

Figure 7. Simulated and TCLab responses using Cohen-Coon settings.

The PSO optimization simulations were carried out in MATLAB®/SIMULINK®. The PSO settings were defined for all the subsequent PID controller tuning simulations as follows:

- A swarm with size m = 30 particles.
- Each simulation was run for 70 iterations.
- The search intervals used for the controller gains K_p, T_i, and T_d are: [0.1 20], [10 s 150 s], and [1 s 12 s], respectively. The tuning gains obtained with classical tuning rules can be used (e.g., CC) to define the search interval. The interval used for the integral gain was widened to see if the PSO converged for slower TCLab control system responses.
- The inertia weight for initial and final values were ω_{init} = 0.7 and ω_{fin} = 0.4. These values were deemed appropriate considering the number of iterations used, as it can be observed from Figure 8.

In this figure, four different inertia weight intervals $[\omega_{init}, \omega_{fin}]$ were considered: (a) [0.9, 0.2], (b) [0.9, 0.4], [0.7, 0.4], and [0.4, 0.4].

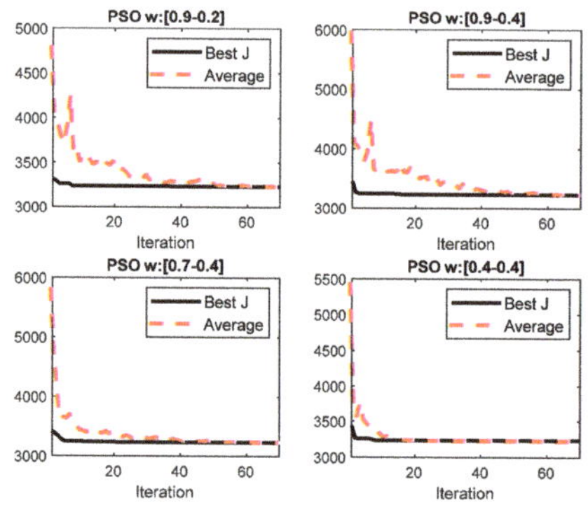

Figure 8. PSO evolution of the best integral absolute error (IAE) values and average swarm IAE values corresponding to four different inertia weight variation settings.

Note that the conditions used for conducting the PSO optimization PID tuning tests differ from the ones used to perform the FOPTD model identification, particularly in the swarm size and higher iteration number. The PSO convergence rate for the PID optimization was found to be slower than the FOPTD model identification case.

The optimizations consider a step response applied at the reference input at $t = 0$ s assuming an ambient temperature of 18 °C and a set-point of 60 °C. The simulation time was 600 s, and a load disturbance with amplitude −40% was applied at instant $t = 300$ s. The optimization test result considering just the minimization of IAE is presented in Figures 9–11. In this case, the gains converged to the following values: $K_p = 13.83$, $T_i = 39.54$ s, and $T_d = 7.14$ s. Figure 9 presents the SPT and LDR responses in the top plot and the control signal in the lower plot. As it can be observed, the system tracks well with no overshoot and with an acceptable load disturbance rejection.

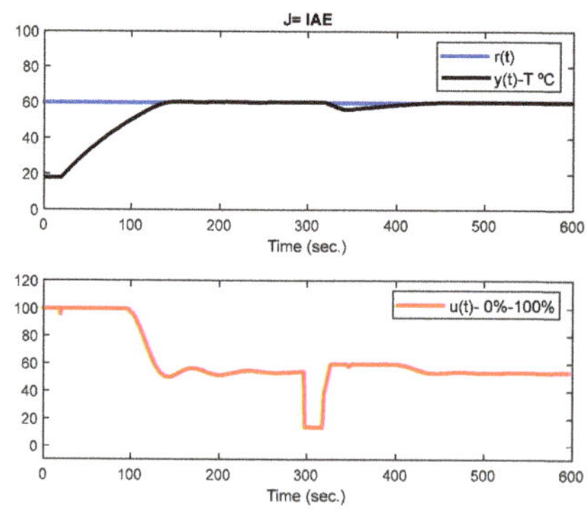

Figure 9. TCLab set-point tracking (SPT) and disturbance rejection (DR) simulated responses with a PID controller optimized with the PSO algorithm (J = IAE).

Figure 10 shows the best PID gains evolution over 70 iterations. It is possible to observe from Figure 10 that the parameters converged to values are very close to the final ones by iteration 40. Figure 11 presents the evolution obtained with the best swarm cost value, IAE, and the average of the entire swarm particles cost function values. As observed, there is a convergence between the average and the best values in between iterations 40 and 50. These results clearly indicate that 70 iterations are an adequate stopping criterion, in this case.

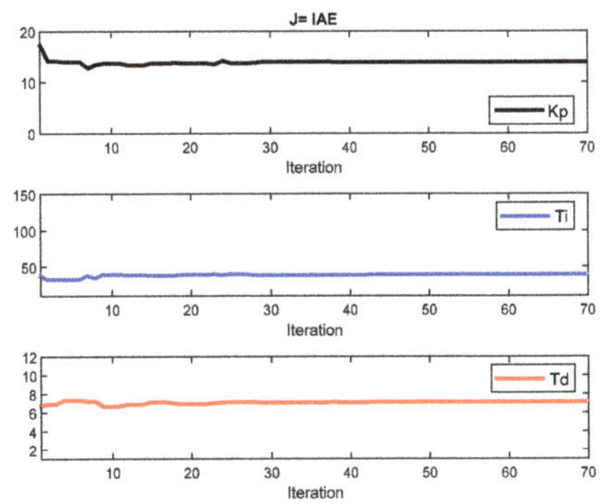

Figure 10. PSO evolution of the PID controller gains resulting in the final values used in the system response of Figure 9.

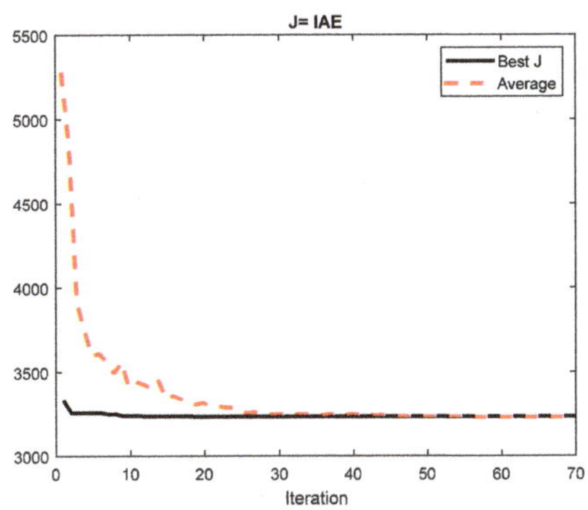

Figure 11. PSO evolution of the best IAE value and average swarm IAE values corresponding to the test in Figures 9 and 10.

After comparing the relative simulated values obtained for IAE and TV, the following cases were considered:

- I: $\alpha = 1$ and $\beta = 0$.
- II: $\alpha = 1$ and $\beta = 70$.
- III: $\alpha = 1$ and $\beta = 80$.
- IV: $\alpha = 1$ and $\beta = 100$.

Notice that this selection was done based on several tests performed on the TCLab. For cases with an additive compromise function, more relevance was given to TV over IAE. Before presenting the TCLab PSO results, a performance comparison with another swarm optimization algorithm, the original Grey Wolf Optimization (GWO) [18], is presented. The GWO was compared with the classical PSO in a benchmark test suit when it was proposed [18]. The GWO swarm size, number of iterations, and PID controller gains interval are the same as the ones used for the PSO. The GWO algorithm uses a heuristic parameter, represented by a [18] responsible for the tradeoff swarm exploration and exploitation. The GWO parameter a is a linearly decreasing vector from $a_{max} = 2$ to $a_{min} = 0$ through the evolutionary process. The simulation results are presented in Table 2 for cases I and II. The best controller gains from an 11 run set converged to very similar IAE and TV values. This indicates that with the appropriate settings the classical PSO performs, as well as the classical GWO for this control case study. More information regarding a comparison between PSO, GWO, and GSA for the PID control is reported in [19].

Table 2. TCLab simulation results obtained with the PSO and grey wolf optimization (GWO) algorithms. The best results are represented in bold.

Method	*Kp*	*Ti* (s)	*Td* (s)	*IAE*	*TV*
PSO-I	13.83	39.54	7.14	4442.0	**2843.8**
GWO-I	13.82	39.49	7.13	**4441.8**	**2843.8**
PSO-II	9.64	49.29	5.70	**3831.4**	2444.7
GWO-II	10.40	47.55	6.08	3834.4	**2429.0**
PSO-III	10.36	47.70	6.05	4438.5	**2921.4**
GWO-III	10.23	47.97	5.97	**4437.5**	2924.8
PSO-IV	9.53	49.69	5.65	**4430.7**	2936.6
GWO-IV	10.17	48.14	5.93	4436.8	**2927.0**

A compilation of the results with the classic tuning rules and PSO optimization is presented in Table 3. The best PSO gains presented in Table 2 are replicated in Table 3 for comparison. Note also that as the TCLab tests were performed in different day periods, the initial starting ambient temperature varies slightly from test to test. To see if this influenced the results analysis, the IAE and TV values considered only the period [95–620 s] (see Table 4). This interval starting point corresponds to 5 s prior to the second step change at t = 100 s where all the temperature steady-state values are approximately equal to 25 °C.

Table 3. TCLab physical results obtained for the several methods considering the entire test time. The best results are represented in bold.

Method	*Kp*	*Ti* (s)	*Td* (s)	*IAE*	*TV*
CC	13.47	45.99	7.00	2389.4	2962.2
AMIGO	4.69	72.96	9.48	3004.9	1972.3
PSO-I	13.83	39.54	7.14	**2273.6**	3600.4
PSO-II	9.64	49.29	5.70	2298.5	2216.7
PSO-III	10.36	47.70	6.05	2392.5	2164.8
PSO-IV	9.53	49.69	5.65	2482.7	**1888.4**

Table 4. TCLab physical results obtained for the several methods considering the time interval [95–620 s]. The best results are represented in bold.

Method	*IAE*	*TV*
CC	2300.9	2647.9
AMIGO	2953.5	1860.6
PSO-I	**2235.8**	3594.1
PSO-II	2259.9	2102.6
PSO-III	2351.1	2065.8
PSO-IV	2442.0	**1788.8**

The results obtained with the AMIGO PID settings: $Kp = 4.69$, $Ti = 72.94$ s, and $Td = 9.48$ s are presented in Figure 12. As it can be observed, the response presents a pronounced overshoot and worse load rejection disturbance compared with the CC response. The TV value obtained with the AMIGO settings is clearly better than the one obtained with CC (see Tables 3 and 4).

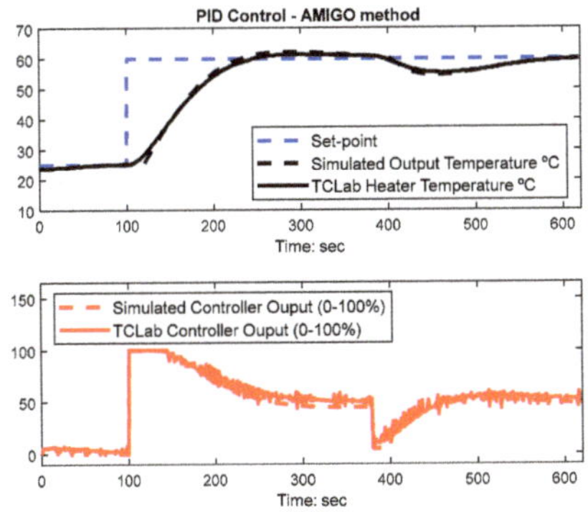

Figure 12. Simulated and TCLab responses using AMIGO settings.

The results obtained for PSO with just the ITAE minimization criterion (PSO-I) are presented in Figure 13. As it can be seen from Tables 3 and 4, only minimizing the IAE results in the best possible IAE value and worse TV than with the CC and AMIGO methods.

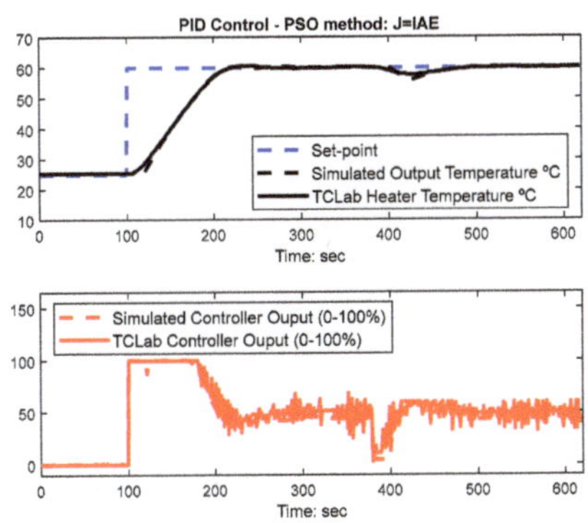

Figure 13. Simulated and TCLab responses using PSO, case I (J = IAE).

The results obtained by using cost functions (8) with $\alpha = 1$ and $\beta = 70$ (PSO-II), $\alpha = 1$ and $\beta = 80$ (PSO-III), $\alpha = 1$ and $\beta = 100$ (PSO-IV), are presented respectively in Figures 14–16.

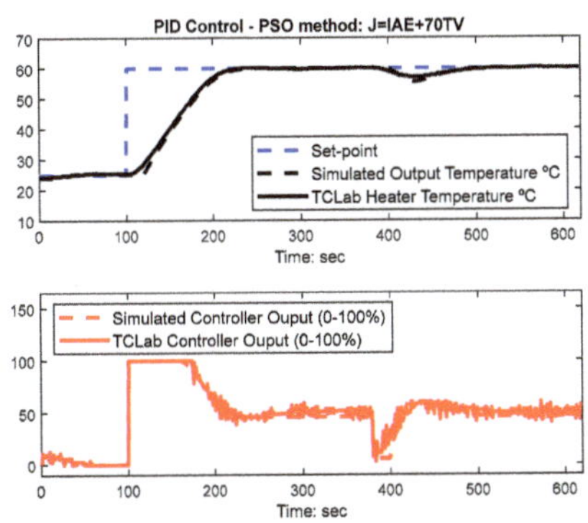

Figure 14. Simulated and TCLab responses using PSO, case II (J = IAE + 70 TV).

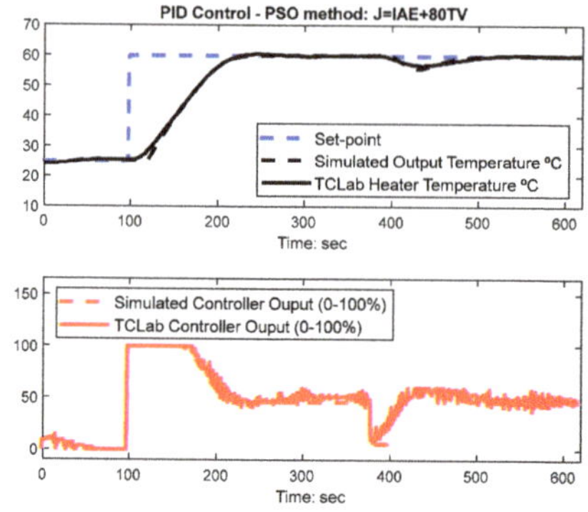

Figure 15. Simulated and TCLab responses using PSO, case III (J = IAE + 80 TV).

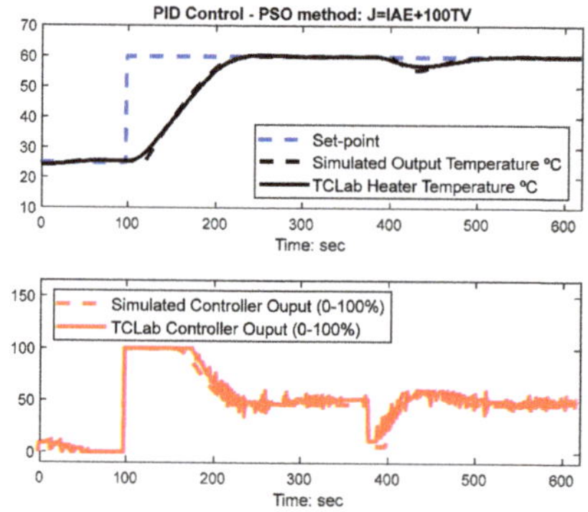

Figure 16. Simulated and TCLab responses using PSO, case IV (J = IAE + 100 TV).

Observing the values obtained with the four PSO cases, the IAE value increases as expected. With the increase of the TV factor, β in the cost function, the corresponding TV value decreases. Depending on the TV requirements, it is possible to obtain better results for both IAE and TV compared with the CC and AMIGO tuning rules, as expected.

Figure 17 presents the overlap responses obtained with the TCLab regarding the following methods: CC, AMIGO, and PSO-IV. As it can be observed from the tracking response in the top plot, CC gains perform slightly better than the PSO-IV gains and these two are much better than AMIGO. However, regarding the control signals in the bottom plot, the optimized PSO gains result in a less aggressive and less irregular control signal variation than CC and AMIGO.

The results also consider an output disturbance applied at instant t = 600 s, as presented in Figure 18. In this case, the disturbance was caused by blowing with a straw near the transistor 1 for 5 s, to decrease the temperature. As it can be observed, the response also tracks well from the output disturbance.

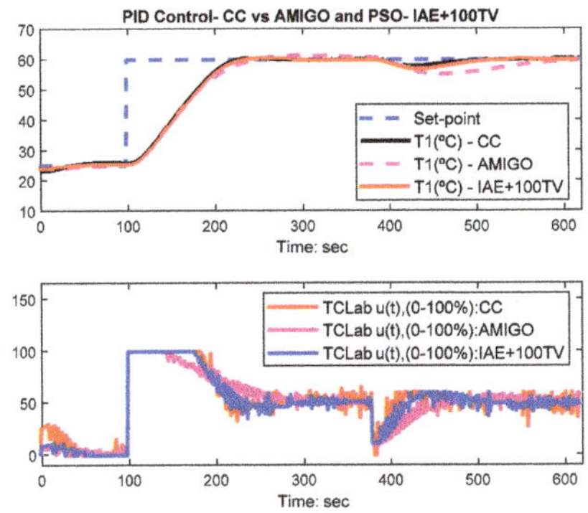

Figure 17. TCLab responses overlapping the responses for Cohen-Coon (CC), AMIGO, and PSO, case IV (J = IAE + 100 TV).

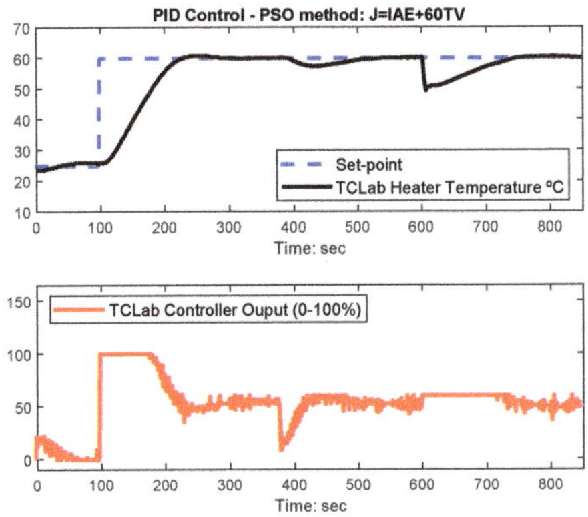

Figure 18. TCLab responses using PSO (J = IAE + 60 TV). Output disturbance considered at t ≅ 600 s.

6. Conclusions

The design of Proportional, Integral, and Derivative controllers (PID) using a technique based on the particle swarm optimization algorithm is presented in this work. Classical PID tuning rules are dependent of the system dynamics and PID controller structure. Moreover, the PID settings in most methods differ according to the specific control design objective (e.g., set-point tracking or load disturbance rejection). The proposed PSO-based PID controller optimization technique holds the following characteristics which are of great practical interest: (a) To be easily implemented in digital industrial computers/microcontrollers; (b) to cope simultaneously with several control design criteria; (c) can be applied to control systems independently of system dynamics; and (d) rely in a reduced number of heuristic parameters, which can be easily adjusted for control purposes.

The merits of the proposed design technique lie within:

- The use of an additive compromise cost function involving the minimization of two performance criteria: The Integral Absolute Error (IAE) and the Total Variation (TV). By using a simple compromise cost function, the proposed PID controller design technique simultaneously considers the following major control design criteria: Set-point tracking, load disturbance rejection, and control signal variation.
- By using the proposed cost function, it was shown that it is quite simple to select the PSO heuristic parameters with a special emphasis on the inertia weight decay.

- The PSO was used to both perform the TCLab system identification as well as PID controller tuning. In both optimization problems, the proposed technique blends quite well with classical design techniques. One of the problems with the use of optimization algorithms in practical problems is to define appropriate decision variable search intervals. The identification of a first-order plus time delay model with the PSO technique uses the two-point step response method [39,40] to define the model parameter search space. The Cohen-Coon PID controller tuning rules were used to define the controller gain search space.
- The PSO results for the specific TCLab control case performs as well as a much more recently introduced metaheuristic: The grey wolf optimization.
- The proposed technique was tested with an Arduino Temperature Control Laboratory (TCLab) and compared with well-established PID tuning methods. Both the simulation as well as physical TCLab results were presented to provide evidence of improved control performance. The results show a good agreement between the simulation and measured results to validate the dynamic model identified with PSO.
- The TCLab Arduino kit was introduced as a simple to use portable laboratory to test simulation results obtained with the proposed PSO-based technique. The same device has great potential to test other optimization and artificial intelligence-based techniques.

Author Contributions: P.B.d.M.O. was involved in the proposed technique conceptualization, validation, TCLab software, and physical experiments; J.D.H. was involved in the TCLab Arduino kit experiment methodology, software, and prosed technique results analysis; E.J.S.P. was involved in the proposed technique validation. All authors were involved in the writing of the original draft preparation, review, and editing. All authors have read and agreed to the published version of the manuscript.

Funding: This research received no external funding.

Conflicts of Interest: The authors declare no conflict of interest.

References

1. Starr, K.; Bauer, M.; Horch, A. An Industry Sponsored Video Course for Control Engineering Practitioners. *IFAC PapersOnLine* **2015**, *48*, 59–64. [CrossRef]
2. PID-18. In Proceedings of the 3rd IFAC Conference on Advances in Proportional-Integral-Derivative Control, Ghent, Belgium, 9–11 May 2018; Available online: https://www.sciencedirect.com/journal/ifac-papersonline (accessed on 20 October 2020).
3. Ziegler, J.G.; Nichols, N.B. Optimum settings for automatic controllers. *Trans. ASME* **1942**, *31*, 759–768. [CrossRef]
4. Peng, Y.; Vrančić, D.; Hanus, R. Anti-windup, bumpless, and conditioned transfer techniques for PID controllers. *IEEE Control Syst. Mag.* **1996**, *16*, 4. [CrossRef]
5. Åström, K.J.; Hägglund, T. *Advanced PID Control*; ISA: Research Triangle Park, NC, USA, 2006; ISBN 1-55617-942-1.
6. Seborg, D.; Edgar, T.F.; Mellichamp, D.A.; Doyle, F.J., III. *Process Dynamics and Control*, 4th ed.; Wiley: Hoboken, NJ, USA, 2017; ISBN 978-1-119-28591-5.
7. Visioli, A.; Vilanova, R. *PID Control in the Third Millennium: Lessons Learned and New Approaches*; Springer: Berlin/Heidelberg, Germany, 2012; ISBN 978-1-4471-2424-5.
8. Rossiter, A.; Zakova, K.; Huba, M.; Serbezov, A.; Visioli, A. A First Course in Feedback, Dynamics and Control: Findings from 2019 Online Survey of the International Control Community. In Proceedings of the IFAC-2020 World Congress, Berlin, Germany, 11–17 July 2020.
9. Holland, J.H. *Adaptation in Natural and Artificial Systems*; The University of Michigan Press: Ann Arbor, MI, USA, 1975.
10. Dorigo, M.; Maniezzo, V.; Colorni, A. Ant system: Optimization by a colony of cooperating agents. *IEEE Trans. Syst. Man Cybern.* **1996**, *26*, 29–41. [CrossRef]
11. Koza, J.R. *Genetic Programming: A Paradigm for Breeding Populations of Computers Programs to Solve Problems*; Technical Report STAN-CS-90-1314; Stanford University: Stanford, CA, USA, 1990.

12. Storn, R.; Price, K.V. Differential Evolution—A Simple and Efficient Adaptive Scheme for Global Optimization over Continuous Spaces. *J. Global Optim.* **1997**, *11*, 341–359. [CrossRef]
13. Kennedy, J.; Eberhart, R.C. Particle swarm optimization. In Proceedings of the International Conference on Neural Networks, Perth, WA, Australia, 27 November–1 December 1995; pp. 1942–1948.
14. Yang, X.-S.; Deb, S. Cuckoo Search via Lévy Flights. In Proceedings of the World Congress on Nature & Biologically Inspired Computing (NaBIC), Coimbatore, India, 8–10 February 2009; pp. 210–214.
15. Yang, X.-S. Firefly Algorithm, Lévy Flights and Global Optimization. In *Research and Development in Intelligent Systems XXVI*; Bramer, M., Ellis, R., Petridis, M., Eds.; Springer: London, UK, 2010. [CrossRef]
16. Krishnanand, K.N.; Ghose, D. Glowworm swarm based optimization algorithm for multimodal functions with collective robotics applications. *Multiagent Grid Syst. Int. J.* **2006**, *2*, 209–222. [CrossRef]
17. Rashedi, E.; Nezamabadi-Pour, H.; Saryazdi, S. GSA: A Gravitational Search Algorithm. *Inf. Sci.* **2009**, *179*, 2232–2248. [CrossRef]
18. Seyedali, M.; Mohammad, S.M.; Lewis, A. Grey wolf optimizer. *Adv. Eng. Softw.* **2014**, *69*, 46–61.
19. Moura Oliveira, P.B.; Vrancic, D.E. Grey Wolf, Gravitational Search and Particle Swarm Optimizers: A Comparison for PID Controller Design. In *CONTROLO 2016*; Lecture Notes in Electrical Engineering; Springer: Guimarães, Portugal, 2016; Volume 402, pp. 239–249. [CrossRef]
20. De Moura Oliveira, P.B.; Freire, H.; Solteiro Pires, E.J. Grey Wolf Optimization for PID Controller Design with Prescribed Robustness Margins. *Soft Comput.* **2016**, *20*, 4243–4255. [CrossRef]
21. Rathore, N.S.; Singh, V.P.; Kumar, B. Controller design for Doha water treatment plant using grey wolf optimization. *J. Intell. Fuzzy Syst.* **2018**, *35*, 5329–5336. [CrossRef]
22. Gupta, S.; Singh, V.P.; Singh, S.P.; Prakash, T.; Rathore, N.S. Elephant herding optimization based PID controller tuning. *Int. J. Adv. Technol. Eng. Explor.* **2016**, *3*, 194. [CrossRef]
23. Jones, A.H.; De Moura Oliveira, P.B. Genetic Auto-Tuning of PID Controllers. In Proceedings of the First IEE Conference on Genetic Algorithms in Engineering Systems: Innovations and Applications (GALESIA'95), Sheffield, UK, 12–14 September 1995; Volume 414, pp. 141–145.
24. Fleming, P.J.; Purshouse, R.C. Evolutionary algorithms in control systems engineering: A survey. *Control Eng. Pract.* **2002**, *10*, 1223–1241. [CrossRef]
25. Coelho, P.C.; De Moura Oliveira, P.B.; Boaventura Cunha, J. Greenhouse Air Temperature Control using the Particle Swarm Optimisation Algorithm. *Comput. Electron. Agric.* **2005**, *49*, 330–344. [CrossRef]
26. Moura Oliveira, P.B.; Solteiro Pires, E.J.; Boaventura Cunha, J.; Pinho, T.M. Review of Nature and Biologically Inspired Metaheuristics for Greenhouse Environment Control. *Trans. Inst. Meas. Control* **2020**, *42*, 2338–2358. [CrossRef]
27. Fan Shu-Kai, S.; Zahara, E. A hybrid simplex search and particle swarm optimization for unconstrained optimization. *Eur. J. Oper. Res.* **2007**, *181*, 527–548.
28. Fan, S.-K.S.; Jen, C.-H. An Enhanced Partial Search to Particle Swarm Optimization for Unconstrained Optimization. *Mathematics* **2019**, *7*, 357. [CrossRef]
29. Irigoyen, E.; Larzabal, E.; Priego, R. Low-cost platforms used in Control Education: An educational case stud. In Proceedings of the 10th IFAC Symposium Advances in Control Education, Sheffield, UK, 28–30 August 2013.
30. Reguera, V.; García, D.; Domínguez, M.; Prada, M.A.; Alonso, S. A low-cost open source hardware in control education. Case study: Arduino-Feedback MS-150. *IFAC PapersOnLine* **2015**, *48*, 117–122. [CrossRef]
31. McLoone, S.C.; Maloco, J. A Cost-effective Hardware-based Laboratory Solution for Demonstrating PID Control. In Proceedings of the UKACC 11th International Conference on Control (CONTROL), Belfast, UK, 31 August–2 September 2016.
32. Prima, E.C.; Saeful, K.; Utarib, S.; Ramdanib, R.; Putri, E.R.; Darmawatib, S.M. Heat Transfer Lab Kit using Temperature Sensor based Arduino TM for Educational Purpose. *Procedia Eng.* **2017**, *170*, 536–540. [CrossRef]
33. Rossiter, J.A.; Pope, S.A.; Jones, B.L.; Hedengren, J.D. Evaluation and demonstration of take-home laboratory kit, Invited Session: Demonstration and poster session. In Proceedings of the 12th IFAC Symposium on Advances in Control Education, Philadelphia, PA, USA, 7–9 July 2019.
34. Juchem, J.; Chevalier, A.; Dekemele, K.; Loccu, M. Active learning in control education: A pocket-size PI(D) setup. In Proceedings of the IFAC-2020 World Congress, Berlin, Germany, 11–17 July 2020.
35. Hedengren, J.D. Temperature Control Lab Kit. Available online: https://apmonitor.com/heat.htm (accessed on 20 September 2020).

36. Hedengren, J.D.; Martin, R.A.; Kantor, J.C.; Reuel, N. Temperature Control Lab for Dynamics and Control. In Proceedings of the AIChE Annual Meeting, Orlando, FL, USA, 10–15 November 2019.
37. Park, J.; Patterson, C.; Kelly, J.; Hedengren, J.D. Closed-Loop PID Re-Tuning in a Digital Twin by Re-Playing Past Setpoint and Load Disturbance Data. In Proceedings of the AIChE Spring Meeting, New Orleans, LA, USA, 31 March–4 April 2019.
38. Park, J.; Martin, R.A.; Kelly, J.D.; Hedengren, J.D. Benchmark Temperature Microcontroller for Process Dynamics and Control. *J. Comp. Chem. Eng.* **2020**, *135*, 6. [CrossRef]
39. Moura Oliveira, P.B.; Hedengren, J.D. An APMonitor Temperature Lab PID Control Experiment for Undergraduate Students. In Proceedings of the 24th IEEE International Conference on Emerging Technologies and Factory Automation (ETFA), Zaragoza, Spain, 10–13 September 2019; pp. 790–797.
40. Moura Oliveira, P.B.; Hedengren, J.D.; Rossiter, J.A. Introducing Digital Controllers to Undergraduate Students using the TCLab Arduino Kit. In Proceedings of the IFAC-2020 World Congress, Berlin, Germany, 11–17 July 2020.
41. Freire, H.; Moura Oliveira, P.B.; Pires, E.J. From Single to Many-objective PID Controller Design using Particle Swarm Optimization. *Int. J. Control Autom. Syst.* **2017**, *15*, 918–932. [CrossRef]
42. Bansal, J.C.; Singh, P.K.; Saraswat, M.; Verma, A.; Jadon, S.S.; Abraham, A. Inertia Weight Strategies in Particle Swarm Optimization. In Proceedings of the IEEE Third World Congress on Nature and Biologically Inspired Computing, Salamanca, Spain, 19–21 October 2011; pp. 633–640.
43. Cohen, G.H.; Coon, G.A. Theoretical consideration of retarded control. *Trans. ASME* **1953**, *75*, 827–834.

algorithms

MDPI

Article

A NARX Model Reference Adaptive Control Scheme: Improved Disturbance Rejection Fractional-Order PID Control of an Experimental Magnetic Levitation System

Hossein Alimohammadi [1], Baris Baykant Alagoz [2], Aleksei Tepljakov [1,*], Kristina Vassiljeva [1] and Eduard Petlenkov [1]

1 Department of Computer Systems, Tallinn University of Technology, 12618 Tallinn, Estonia; hossein.alimohammadi@taltech.ee (H.A.); kristina.vassiljeva@taltech.ee (K.V.); eduard.petlenkov@taltech.ee (E.P.)

2 Department of Computer Engineering, Inonu University, 42280 Malatya, Turkey; baykant.alagoz@inonu.edu.tr

* Correspondence: aleksei.tepljakov@taltech.ee

Received: 15 July 2020; Accepted: 14 August 2020; Published: 18 August 2020

Abstract: Real control systems require robust control performance to deal with unpredictable and altering operating conditions of real-world systems. Improvement of disturbance rejection control performance should be considered as one of the essential control objectives in practical control system design tasks. This study presents a multi-loop Model Reference Adaptive Control (MRAC) scheme that leverages a nonlinear autoregressive neural network with external inputs (NARX) model in as the reference model. Authors observed that the performance of multi-loop MRAC-fractional-order proportional integral derivative (FOPID) control with MIT rule largely depends on the capability of the reference model to represent leading closed-loop dynamics of the experimental ML system. As such, the NARX model is used to represent disturbance-free dynamical behavior of PID control loop. It is remarkable that the obtained reference model is independent of the tuning of other control loops in the control system. The multi-loop MRAC-FOPID control structure detects impacts of disturbance incidents on control performance of the closed-loop FOPID control system and adapts the response of the FOPID control system to reduce the negative effects of the additive input disturbance. This multi-loop control structure deploys two specialized control loops: an inner loop, which is the closed-loop FOPID control system for stability and set-point control, and an outer loop, which involves a NARX reference model and an MIT rule to increase the adaptation ability of the system. Thus, the two-loop MRAC structure allows improvement of disturbance rejection performance without deteriorating precise set-point control and stability characteristics of the FOPID control loop. This is an important benefit of this control structure. To demonstrate disturbance rejection performance improvements of the proposed multi-loop MRAC-FOPID control with NARX model, an experimental study is conducted for disturbance rejection control of magnetic levitation test setup in the laboratory. Simulation and experimental results indicate an improvement of disturbance rejection performance.

Keywords: multi-loop model reference control; PID controllers; FOPID retuning control; disturbance rejection control

1. Introduction

Real-life control systems are subject to unpredictable disturbances that may severely decrease control performance. Therefore, control systems, which operate in real-world conditions, should be designed according to certain robustness criteria to deal with impacts of disturbance incidents on the control system performance. A major weakness of a typical control system is the dependence of the controller on a mathematical formulation of the plant—the so called model-based control design process. For consideration of disturbance dynamics, classical controller tuning requires modeling of disturbance incidences in order to ensure that the impact of disturbances on the system's performance is minimal. However, since disturbances are of unpredictable and uncertain character in many real-world applications, a classical controller may not respond optimally at the onset of a disturbance signal in a control loop. Another factor, which limits inherent disturbance rejection control performance of classical control loops, is the tradeoff between disturbance rejection control and set-point tracking performance. The controller tuning for higher disturbance rejection performance results in deterioration of the tracking performance [1]. Intelligent control techniques should be used to solve these problems. These techniques can detect disturbance incidences and adapt the control system response to reduce the negative impacts of disturbances on the control performance. The motivation for the current study comes from this requirement. In this manner, a two-loop control structure that involves specialized loops is considered: a fractional-order proportional integral derivative (FOPID) control loop for improved stability and set-point control performance, and a Model Reference Adaptive Control (MRAC) loop for improvement of disturbance rejection performance. These specialized control loops work in harness to increase disturbance rejection capability of the control system while preserving set-point tracking performance.

MRAC is one of the widely used approaches for adaptive control of systems [2–4]. The MRAC structure uses a reference model that reinforces the control system according to a predefined reference model. Here, the reference model describes a desired response of the control system, and the control system is designated to resemble responses of the reference model [2]. In the current study, the outer loop is implemented by using Massachusetts Institute of Technology (MIT) rule. The MIT rule is a straightforward technique for performing the MRAC [4–7] process. The MIT rule is essentially based on the decrease of difference between responses of the reference model and the control system by implementing a gradient descent method in adaptation loop (the outer loop). Direct use of the MIT rule for control action brings out some stability concerns because the control action of the MIT rule is very sensitive to output amplitude of the reference model [5]. This shortcoming is a major reason for the development of multi-loop MRAC-FOPID control structures, where the MRAC with MIT (outer loop) rule only activates in case of disturbance incidents. The set-point control and system stability are achieved by the closed-loop control system (the inner loop) [8–11].

On the other hand, fractional control has attracted significant interest in the past few decades [12–14]. Several works have reported robust control performance improvements by using fractional-order controllers [15–18]. In the current study, the FOPID controller is implemented by using a retuning FOPID controller method that was suggested as an effective approach to implementing FOPID controllers while keeping original PID control loops intact [17,18]. The retuning FOPID control loop is used in case of this work for set-point control of the experimental magnetic levitation (ML) system and this loop is nestled into the inner loop of the proposed multi-loop MRAC-FOPID control structure [9]. The outer loop with the MIT rule encloses the inner loop and it is called the adaptation loop. The adaptation loop (the outer loop) is designed to detect the disturbance incidents and perform an adaptation of the inner loop in order to decrease negative effects of disturbance on the control performance. An experimental study of a multi-loop MRAC-FOPID control structure was carried out for magnetic levitation (ML) control problem by Tepljakov et al. [9]. Authors observed that the performance of multi-loop MRAC-FOPID control with MIT rule largely depends on capability of the reference model to represent leading closed-loop dynamics of the experimental ML system. In [9], a linear model of an experimental ML system was used to obtain the transfer function of the closed-loop retuning FOPID control system. This function was used as the reference model. This modeling approach produces two drawbacks for the multi-loop MRAC-FOPID control structure.

i. Linearization of a nonlinear ML system makes the reference model valid at the modeling conditions such as operating temperature, sphere motion ranges etc. This limitation decreases the accuracy of the reference model.
ii. Identification of the inner loop is an initial process and this static reference model is not adaptive for changes in condition and behavior of the real systems.

In the current study, to deal with these drawbacks, nonlinear autoregressive neural network with external inputs (NARX) modeling is employed to obtain a more accurate reference model. The use of neural networks in system identification allows the utilization of machine learning techniques in control, and this makes the multi-loop MRAC-FOPID control structure with the NARX model more adaptive compared to previous configurations. Recurrent neural networks have been effectively used for intelligent control and online system modeling [19–24]. Autoregressive neural network models can online learn the dynamical responses of linear and nonlinear systems from sampled input and output data [21–24]. Therefore, it gains significant flexibility for multi-loop MRAC-FOPID control structures to employ in real applications. NARX modeling can automatically learn responses of a wide range of system models as a black-box model and this reduces the need for guessing the model structure prior to model identification. The reference model is automatically identified from data that are captured from the input and output of controlled systems online. This point is an important contribution of this study to facilitate the use of multi-loop MRAC-FOPID control structures in practical control applications.

Preliminary studies on neural network control have discussed the use of ANNs in control practice and neural control concepts in [25,26]. Nonlinearity in the behavior of real systems is one of the major factors that complicates the control design task and reduces practical control performance [27]. The static linear control based on linearized models may not always yield satisfactory results for control of nonlinear systems because of unpredictable alterations in operation conditions in the real world systems. Several nonlinear control methods have been proposed to address control problems of nonlinear systems in [27]. However today's intelligent system paradigm requires highly adaptive, model-free control systems. ANNs have emerged as a possible candidate for evolution of control systems to meet these requirements [28].

For magnetic levitation control problems, several control approaches have been proposed in the literature. In some research, neural networks were successfully used. For instance, a recurrent neural network-based adaptive optimal backstepping strategy was proposed—in the work, authors used a dynamic model of the recurrent neural network (RNN) to solve the constrained optimization problem to achieve improved control performance [29]. In another work using ANNs [30], authors suggested a radial basis function (RBF) network to perform control and another RBF network for model identification of the system. As a third component, a PID controller was used for the stabilization of this three component control system. In addition to neural control approaches, some recent works also demonstrated the use of other control approaches such as observer-based adaptive control [31] and the Takagi–Sugeno fuzzy controller [32] a for magnetic levitation control problem. However, these works have not purposely focused on disturbance rejection control performance of magnetic levitation control system.. The current study specifically addresses disturbance rejection control of a magnetic levitation system and experimentally validates performance improvements.

To improve the performance of classical integer-order MRAC with the MIT rule, fractional-order MRAC approaches have been suggested in several works [3]. Vinagre et al. have introduced fractional-order adjustment rules and use of fractional order reference models. They indicated the improvement in adaptation performance by using fractional order MIT rule [3]. Then, Ladaci et al. suggested the use of fractional-order derivative at the controlled system output as a feedback to fractional-order MIT rule [33]. This modification was shown to improve noise performance of fractional-order MIT rule in simulation examples [33]. Experimental validation of fractional order MIT rule was demonstrated for an experimental rotor control system [7]. These methods are named direct MRAC because the MIT rule is directly used to control a class of plant functions.

The main difference of the multi-loop MRAC from the direct MRAC is that the MIT rule in multi-loop MRAC does not account for stability and control actions of the plant. Stability and optimal control are ensured by an inner control loop, which is commonly an optimal closed-loop control system. Several fractional-order adaptive control approaches have been proposed to improve control performance of fractional order MRAC systems such as a composite model reference adaptive control approach using tracing and prediction errors [34], fractional adaptive controller method with controller parameter adjustment for automatic voltage regulator [35], MRAC with Bode ideal loop reference model [36].

The motivation behind the present research effort has thus been established. The contribution of this research work can be stated as follows: a multiloop control method is proposed based on the FOPID retuning approach and computational intelligence in the form of an artificial neural network-based reference model used in the MRAC scheme for endowing the complete control system with greater robustness without deteriorating set-point tracking performance. The complete control system is initially simulated using a model of a magnetic levitation system and then real-time experiments are carried out confirming the improvement in the performance of the control system.

Organization of the study is as follows: Section 2 provides a theoretical background for all control methods considered in this study. In Section 3, the approaches to modeling the magnetic levitation system are discussed including the first principles approach and the artificial neural network based black box approach. Section 4 introduces multi-loop MRAC-FOPID control with the obtained NARX reference model and presents results stemming from simulation and experimental studies. Finally, conclusions are drawn in Section 5.

2. Theoretical Background and Preliminaries

2.1. Fractional Calculus and Fractional-Order Systems

Fractional calculus introduces non-integer order operator and allows the orders of the operator to be rational, real, and complex numbers. This property offers a more coherent and flexible mathematical modeling approach of real-world system dynamics.

The fractional derivative operator was presented in [12,13] as

$$ {}_aD_t^{\alpha} = \begin{cases} \frac{\mathrm{d}^{\alpha}}{\mathrm{d}t^{\alpha}} & \alpha > 0 \\ 1 & \alpha = 0 \\ \int_{\alpha}^{t} (\mathrm{d}\tau)^{(-\alpha)} & \alpha < 0 \end{cases}, \tag{1} $$

where (generally) $\alpha \in \mathbb{R}$ is the fractional order (noninteger order). For $\alpha > 0$, ${}_aD_t^{\alpha}$ operator expresses a fractional-order derivative, and for $\alpha < 0$, ${}_aD_t^{\alpha}$ operator expresses a fractional-order integral operator. The parameters a and t determine the lower and upper limits of the operation. Several mathematical definitions of fractional-order derivative operator have been proposed over some 300 years of the development of fractional calculus.Widely preferred definitions are the Grünwald–Letnikov definition, the Riemann–Liouville definition and the Caputo definition [12–14].

The Riemann–Liouville definition is preferred for system modeling and control engineering practice because the Laplace transform of this definition can be expressed as s^{α} in the s-domain assuming zero initial conditions. This simplifies fractional-order transfer function modeling. Fractional-order differential equations are written in general form as [12–14]:

$$ a_nD^{\alpha_n}y(t) + a_{n-1}D^{\alpha_{n-1}}y(t) + ... + a_0D^{\alpha_0}y(t) = b_mD^{\beta_m}u(t) + b_{m-1}D^{\beta_{m-1}}u(t) + ... + b_0D^{\beta_0}u(t), \tag{2} $$

where, α_i, β_j stands for fractional-orders of the system model and $\alpha_0 < ..... < \alpha_{n-1} < \alpha_n$ and $\beta_0 < ..... < \beta_{m-1} < \beta_m$ are real numbers. The parameters a_i, b_j are constants and denotes coefficients of time invariant fractional-order system models.

This fractional-order differential equation can model fractional-order dynamics in the time domain. Fractional-order control system design commonly requires transfer function models to perform s-domain design approaches. The transfer function models can be derived by applying a Laplace transform to the fractional-order differential equation model given by Equation (2). Due to providinga Laplace transform, the Riemann–Liouville definition of fractional-order derivative is frequently preferred in control theory and it was defined as

$$ {}_aD_t^\alpha f(t) = \frac{1}{\Gamma(n-\alpha)} \frac{\mathrm{d}^n}{\mathrm{d}t^n} \int_a^t \frac{f(\tau)}{(t-\tau)^{\alpha-n+1}} \mathrm{d}\tau, \tag{3} $$

where $\Gamma(.)$ is Gamma function and $n-1 < \alpha < n$. By assuming zero initial conditions, Laplace transform of Riemann–Liouville fractional derivative is expressed as follows [12,37]:

$$ L\{D^\alpha f(t)\} = s^\alpha F(s) \tag{4} $$

By taking Laplace transform of Equation (2), again assuming zero initial condition, fractional-order transfer functions are written in a general form by [12,37,38]

$$ G(s) = \frac{b_m s^{\beta_m} + b_{m-1} s^{\beta_{m-1}} + \ldots + b_0 s^{\beta_0}}{a_n s^{\alpha_n} + a_{n-1} s^{\alpha_{n-1}} + \ldots + a_0 s^{\alpha_0}} = \frac{\sum_{j=0}^{m} b_j s^{\beta_j}}{\sum_{i=0}^{n} a_i s^{\alpha_i}}. \tag{5} $$

After obtaining the transfer function model, the characteristic equation of fractional-order transfer function can be used for stability analysis. The fractional-order characteristic equation of $G(s)$ is written by

$$ \Delta(s) = \sum_{i=0}^{n} a_i s^{\alpha_i} = 0. \tag{6} $$

Root locus of this characteristic equation is analyzed for stability property evaluation of fractional-order systems [39–42].

In control engineering, FOPID controllers are widely expressed in transfer function form [12]

$$ C(s) = k_p + \frac{k_i}{s^\lambda} + k_d s^\mu,, \tag{7} $$

where parameters k_p,k_d and k_i are controller coefficients, and $\lambda \in R$ and $\mu \in R$ represent fractional-orders of the controller function; if $\lambda = \mu = 1$, then a conventional PID controller is obtained. Fractional-order derivative is not a local operator. For this reason, ideal digital realization of fractional-order elements uses growing computational resources in time as a result of the long memory effect. Therefore, for the practical realization of this controller, approximate fractional-order models are used to implement fractional-order derivative and integral elements [43–46]. In practice, FOPID controllers are commonly implemented by using these approximate fractional-order models.

A major advantage of fractional order derivative and integral element comes from the property that fractional-order elements provide flexibility for tuning of dynamic responses in a range from local to non-local operation. Let us consider this benefit for the closed-loop FOPID control system: Integer order derivative is originally a local operator and it considers very recent changes in the control error while generating a control action. However, fractional order derivative exhibits non-locality in differentiation operation. By adjusting the fractional order of the derivative element, FOPID controller can be tuned to consider all previous changes in control error while generating a control action [3]. Such adjustability option in the locality of fractional order elements can contribute to better modeling of real systems, which presents the property of non-locality in time and space, and to generate a better response to such a real system dynamics in control applications.

In this study, the Oustaloup approximation method, which is widely preferred in fractional-order control system simulations, was utilized in the implementation of FOPID controller in simulation and experimental systems.

To replace existing PID controllers in control loops with FOPID-based dynamics without breaking the existing loop, a retuning FOPID controller structure was proposed for the practical implementation of closed-loop FOPID controllers [17,18]. The transfer function of the retuning FOPID controller is expressed in the form of

$$C(s) = (C_R(s) + 1)C_{PID}(s), \tag{8}$$

where, $C_{PID}(s)$ stands for the existing PID controller. The retuning controller function is found as

$$C_R(s) = \frac{C(s)}{C_{PID}(s)} - 1. \tag{9}$$

Figure 1 depicts the integration of the retuning controller function $C_R(s)$ to an existing PID control loop to realize a FOPID control system.

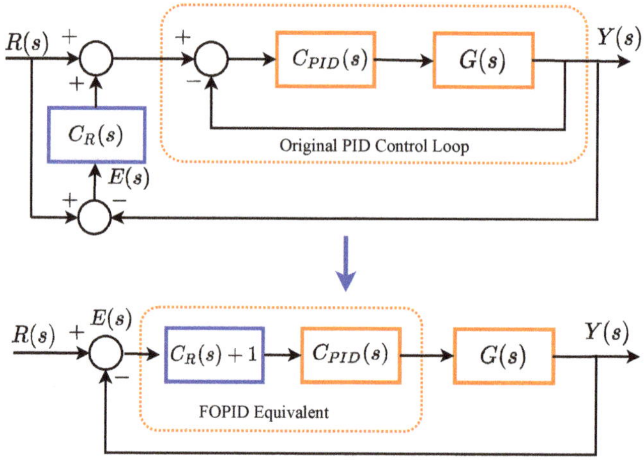

Figure 1. Implementation of the retuning fractional-order proportional integral derivative (FOPID) system by using the PID control loop [9].

This method is of high industrial interest because it allows us to replace existing, suboptimally tuned PID control loops with optimally tuned PID or FOPID controllers essentially without interrupting the control process since the dynamics are introduced through manipulating the original reference signal. This means that the controlled process ideally does not experience any downtime and hence this change in control strategy should not result in additional expenses related to restarting the process after the change of controller has taken place.

As it can be seen from Figure 1, when discussing retuning control, in an ideal scenario one can just speak about the control loop consisting of a FOPID controller and plant in a negative unity feedback configuration because, at least from a purely theoretical perspective, the retuning method ensures that the complete control loop will indeed have the dynamics introduced by the external FOPID controller instead of those stemming from the original PID controller. In other words, it is equivalent to discussing a simpler FOPID control loop rather than the retuning control loop, at least for the sake of the simplicity of exposition.

2.2. Multi-Loop Mrac

Direct use of the MIT rule for control action brings out some instability concerns because the control action of the MIT rule is very sensitive to output amplitude of reference models [5]. This shortcoming is the main motivation for the development of two-loop MRAC-FOPID control structures, where an outer loop, which is an MRAC loop with MIT rule, only performs for adaptation to reference model.

The two-loop MRAC-FOPID control has been proposed to improve the robust control performance of the conventional FOPID control loop [8,9]. For this purpose, an additional outer loop to improve adaptation characteristics of the control system is appended to the FOPID control loop. In other words, the outer loop with a classical integer-order MIT rule encloses the inner loop and it is called the adaptation loop. This simplifies the design task of the outer loop. Thus, stability and set-point control performance are managed by the inner loop that is a FOPID control loop, and the disturbance rejection control to improve robust control performance is provided by the outer loop that is an MRAC loop with MIT rule. The design task of this control structure includes the following basic steps [8,9]:

Step 1: The inner loop (the FOPID control loop) is designed optimally by using any optimal FOPID tuning method. Robust stability and a satisfactory set-point control performance should be achieved by using a suitable optimal controller tuning method.

Step 2: A model of the inner loop is obtained and used as a reference model of MRAC structure.

Step 3: The outer loop is connected to the inner loop according to MIT rule.

The final structure of the two-loop MRAC-FOPID control is shown in Figure 2. In the figure, the adaptation gain θ is multiplied to the reference input r in order to shape the reference input when the response of the reference model $T_m(s)$ diverges from the response of FOPID loop due to disturbance incidents. Therefore the outer loop, which is also known as the adaptation loop, enforces the inner loop, which is called the control loop, to behave similarly to the reference model that describes disturbance-free and optimal responses of the inner loop. Specifically, Step 2 ensures the correct establishment of this mechanism. It should be noted that in this work, Step 2 is modified. This will be clear from the following discussion. Specifically, the reference model is obtained independently from the inner loop.

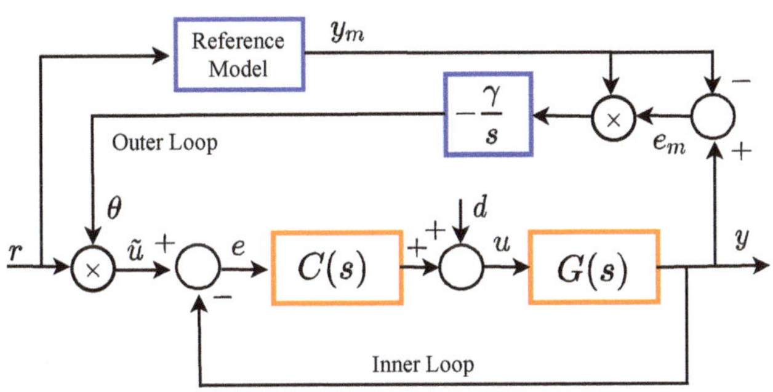

Figure 2. Block diagram of multi-loop Model Reference Adaptive Control (MRAC)-FOPID system [9].

The closed-loop retuning FOPID control system was implemented and the FOPID controller was optimally tuned by using FOMCON toolbox [47].

The theoretical background of MIT rule to implement the outer loop was summarized as follows [8,9]:

The MIT rule works for the minimization of model error that is defined according to difference between outputs of inner loop (FOPID control loop) and the reference model as $e_m = y - y_m$. Therefore, the error function to be minimized is written in form of square of instant model error

$$J = \frac{1}{2} e_m^2. \tag{10}$$

According to Figure 2, to perform adaptation by applying input-shaping technique, the input of closed-loop system is modified as $\tilde{u} = \theta\, r$, where the adaptation gain θ is determined according to the MIT rule. The MIT rule performs the continuous time gradient descent method to minimize the error function, it was expressed for MRAC [2–6] as

$$\frac{d\theta}{dt} = -\gamma \frac{dJ}{d\theta} = -\gamma e_m \frac{de_m}{d\theta}. \tag{11}$$

By considering multi-loop MRAC-FOPID structure in Figure 2, the model error can be expressed as [8,9]

$$e_m = y - y_m = T(s)\theta r - T_m(s) r. \tag{12}$$

Then, the sensitivity derivative $de_m/d\theta$ is written by

$$\frac{de_m}{d\theta} = T(s) r, \tag{13}$$

where the transfer function $T(s)$ represents the current model of the inner loop. When the reference input is substituted with $r = y_m/T_m(s)$ in Equation (13), the sensitivity derivative can be obtained

$$\frac{de_m}{d\theta} = \frac{T(s)}{T_m(s)} y_m. \tag{14}$$

By using the sensitivity derivative in Equation (11), MIT rule for the update of adaptation gain is obtained as

$$\theta = -\gamma \frac{1}{s} \left(\frac{T(s)}{T_m(s)} y_m e_m \right). \tag{15}$$

To investigate the contribution of the adaptation gain θ to system response, let us consider the response of the system in disturbance-free state and in case of a disturbance incident.

(i) Disturbance-free state: In this state, one can assume $T(s) = T_m(s)$ because reference model is configured as mathematical model of optimally tuned inner model. Accordingly, the adaptation gain can be written as,

$$\theta = -\gamma \frac{1}{s} y_m e_m. \tag{16}$$

In the disturbance-free state, outputs of reference model and the inner loop match each other, hence $e_m = 0$. When $e_m = 0$ and $T(s) = T_m(s)$ conditions are used in Equation (12), one obtains $e_m = T_m(s)(\theta - 1) r = 0$. This arithmetically infers the state of $\theta = 1$ because the reference model $T_m(s)$ is non-zero function and the reference signal r is the independent input variable. Therefore, adaptation gain settles the value of $\theta = 1$ in the disturbance-free state. It does not shape the reference input. This property validates that outer loop does not affect control performance of the inner loop in disturbance free case.

(ii) Disturbance state: In case of a disturbance incident, one can assume a perturbation of the system model due to additional dynamics of the additive interference of disturbance. This effect results in differentiation of inner loop model from the reference model, and let us assume $T(s) \neq T_m(s)$. The MIT rule is based on gradient descent perform optimization of the adaptation gain θ to minimize error function and it leads to $\lim_{t \to \infty} e_m(t) \to 0$. Therefore, one can consider the case $e_m = T(s)\theta\, r - T_m(s) r \to 0$, where the adaptation should change $\theta \to \frac{T_m(s)}{T(s)}$ to reach the state $e_m = 0$ in time. When the MIT rule achieves the state $e_m = 0$, response of the reference model and response of the inner loop become the same, and impacts of disturbance incident at the system output are rejected by the change of $\theta \rightarrow \frac{T_m(s)}{T(s)}$.

This analysis reveals the fact that the performance of the multi-loop MRAC-FOPID control system depends on the capability of the reference model to represent the response of the closed-loop FOPID control system. Therefore, identifying an accurate model of the closed-loop FOPID control system is of vital importance. This requirement constitutes the central motivation of this study. Therefore, practical use of NARX modelling for multi-loop MRAC-FOPID control is tested in a relatively difficult control problem that is the disturbance rejection control of a nonlinear, unstable dynamical system, namely the ML system.

To address stability concerns related to this configuration, the relevant discussion is provided next. In particular, it is important to address to components of stability analysis: the theoretical one and the empirical one. In the former case, stability conditions are derived mathematically, in the latter one, experiments are conducted and the results analyzed to ensure that the control system is indeed stable.

Stability and convergence conditions of the multi-loop MRAC-FOPID control system was summarized as follows in [11]:

The reference model and inner loop are designed as a stable system. Therefore, a sufficient condition for stability of the multi-loop MRAC-FOPID control system can be written based on stability and convergence of model error $|e_m| = |y - y_m| < \epsilon \in R$. Assuming that the inner loop and reference model are designed stable, multi-loop MRAC-FOPID control systems are stable and convergent when model error e_m is bounded, the input is bounded, and output is stable and convergent. Therefore, it is sufficient to investigate the convergence condition of model error e_m to show the stability of multi-loop MRAC--FOPID control system.

Lemma 1. *(Zero condition of the model error [11]): For multi-loop MRAC-FOPID control structure, the model error takes a zero value ($e_m = 0$) when the adaptation gain θ is equal to one and the transfer functions of reference model and inner loop are equal ($T(s) = T_m(s)$).*

Proof. By considering the model error given by Equation (12) for multi-loop MRAC-FOPID control structure, the model error $e_m = y - y_m$ can be written for $\theta = 1$, and $T(s) = T_m(s)$ as

$$e_m = T(s)\theta r - T_m(s)r = T(s)r - T_m(s)r = 0. \tag{17}$$

Then, the model error becomes zero. This proves a zero condition of the model error as to be $\theta = 1$ and $T(s) = T_m(s)$. □

Theorem 1. *(Stability of model error [11]): For a stable system models $T(s) = T_m(s)$, multi-loop MRAC--FOPID control system is stable only if the model error function:*

$$e_m = -\frac{s^2}{\gamma}\left(\frac{1 + C(s)G(s)}{C(s)G(s)}\right). \tag{18}$$

expresses a stable function.

Proof. By using the adaptation rule of multi-loop MRAC-FOPID control system that was given by Equation (16) as $\theta = -\gamma\frac{1}{s}y_m e_m$, the model error can be derived as

$$e_m = -\frac{s\theta}{\gamma y_m}. \tag{19}$$

Output of reference model y_m can be written as $y_m = T_m(s)r(s)$. For a set-point control stability, one considers step response of the system. Therefore, the reference input is assumed to be a step function by substituting $r(s) = \frac{1}{s}$ in Equation (19).

$$e_m = -\frac{s^2\theta}{\gamma T_m(s)} \quad (20)$$

According to Lemma 1, a zero value of model error ($e_m = 0$) is possible for the case of $\theta = 1$, and $T(s) = T_m(s)$. By applying this condition, the model error is obtained as

$$e_m = -\frac{s^2}{\gamma T(s)} \quad (21)$$

The function $T(s)$ is transfer function of inner loop that is a closed-loop FOPID controller. Therefore, it can be expressed in the form of $T(s) = \frac{1}{1+C(s)G(s)}$. When $T(s)$ is used in Equation (21), one can write and reach to Equation (18).

The multi-loop MRAC--FOPID control system is stable if the function e_m given Equation (18) is a stable function (check if the all characteristic roots of characteristic equation of e_m lies on the left half plane (LHP)). Then, one can infer that when e_m function is LHP stable function, the amplitude $|e_m| = |y - y_m| < \epsilon \in R$ is bounded. Since the reference model was always configured to be stable function in design, the reference model output is bounded ($|y_m| < \epsilon \in R$) and it leads output of plant is also bounded ($|y| < \delta \in R$), and whole system is stable. □

3. Mathematical Model of the Ml System: From a First Principles Model to a Narx Black Box Approximation

The content of the following section conveys the mathematical modeling of the ML system using two different approaches. First, first-principles modeling is considered; then, the NARX model is developed. It is important to compare the accuracy of both models to ensure that the latter represents a coherent model of the ML system's dynamics.

The Magnetic Levitation System (MLS) is a nonlinear, open-loop unstable, time-varying dynamical system with negligible air friction force. Figure 3 shows a picture of an experimental ML setup that was used for testing the proposed control structure. The objective in this experimental system is to control the terminal voltage of an electromagnet so that the ferromagnetic sphere can be levitated at the desired distance from the magnet. One of the promising applications of ML is the ML transportation systems. Specifically, physically contactless transportation on a guide-way can offer a number of advantages such as lower noise, less friction forces, less costly maintenance, higher efficiency and less emission of exhaust fumes. These advantages make ML transportation a suitable candidate for future transportation possibilities [48,49]. For these reasons, robust control of the ML system introduces a significant control problem that has the potential of producing technological outcomes. Several works have addressed robust control performance of ML systems; for instance, fuzzy logic controllers [50] and FOPID controllers [9,17,51–53] were considered. It is important to stress that disturbance rejection and vibration control are essential problems in control of ML systems.

In order to compare among results including experimental and artificial models, the mathematical modeling of ML systems has been studied.

Magnetic levitator can be divided into two subsystems, mechanical and electrical [50,54]. For the electrical subsystem, it is important to single out electromagnetic coil inductance L (H), and its resistance, R (Ω). The electrical subsystem can be described with the following well-known differential equation:

$$U = R_l i + L\frac{di}{dt}. \quad (22)$$

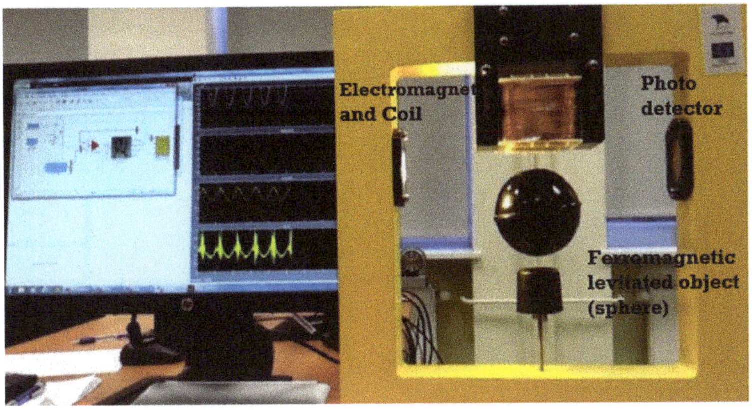

Figure 3. Magnetic Levitation System (MLS) setup in A-Lab.

In order to determine the current in the coil, the resistor R_s is connected in series with the coil. In that way, voltage U_s can be measured across resistor R_s, by using A/D convertor for measuring current i. Now, Equation (22) can be rewritten as:

$$U = (R_i + R_s)i + L\frac{di}{dt}. \tag{23}$$

After applying Laplace transform we obtain:

$$Ge(s) = \frac{I(s)}{V(s)} = \frac{1}{Ls + (R_i + R_s)}. \tag{24}$$

On the other side, modelling mechanical subsystem can be performed simply by defining force F, that represents the result of electromagnet activity to the ball:

$$F = mg - K_f(\frac{i}{x})^2, \tag{25}$$

where m is ball mass, g is gravitation constant, K_f is magnetic force constant which is valid for pair electromagnet and ball, i is the current which flows through electromagnet, x is ball distance from magnet. Using second Newton's law, Equation (25) can be rewritten as:

$$m\frac{d^2x}{dt^2} = mg - K_f(\frac{i}{x})^2. \tag{26}$$

The value of electromagnetic coil current in steady state denoted as I_{ss} can be determined by Equation (26). This current defines a constant of ball position X_{ss} in a steady state case. If we take into consideration that velocity and acceleration are equal to zero in the state, then Equation (26) is given as:

$$I_{ss} = \sqrt{\frac{mg}{K_f}}X_{ss}. \tag{27}$$

Theoretically, regulation of the ball position can be performed by (27). However, external disturbances, uncertainty, system parameter variations, and high nonlinearity degree of the system require a feedback controller installation, which ensures stable control of these mechanical subsystems.

In general, adding linearity into the system must be done first, and after that control process begins. In this report, NARX neural network is used for experimental substitution of the real system of magnetic levitation. If the training process is successfully performed, we are going to get values on network outputs with minimal errors comparing to the output values of the real system. The nonlinear mathematical model of magnetic levitation system is given by the following equations [55]:

$$\dot{x_1} = x_2, \tag{28}$$

$$\dot{x}_2 = -\frac{F_{em}}{m} + g, \tag{29}$$

$$\dot{x}_3 = \frac{1}{f_i(x_1)}(k_i u + c_i - x_3), \tag{30}$$

$$F_{em} = x_3^2 \frac{F_{emP1}}{F_{emP2}} exp(-\frac{x_1}{F_{emP2}}), \tag{31}$$

$$f_i(x_1) = \frac{f_{iP1}}{f_{iP2}} exp(-\frac{x_1}{f_{ip2}}), \tag{32}$$

where x_1, x_2, x_3 are ball position, velocity and current respectively. u is the control signal. The corresponding parameters are given in Table 1.

Table 1. Physical parameters of the ML system.

Parameter	Physical Description	Unit
$m = 5.7100 \times 10^{-2}$	mass of ball	[kg]
$g = 9.81$	gravity constant	$[m/s^2]$
$F_{em} = f(x_1, x_3)$		[N]
$F_{emP1} = 1.7521 \times 10^{-2}$	electromagnetic force	[H]
$F_{emP2} = 5.8231 \times 10^{-3}$	electromagnetic force	[m]
$f_i(x_1)$		[1/s]
$f_{iP1} = 1.4142 \times 10^{-4}$		[ms]
$f_{iP2} = 4.5626 \times 10^{-3}$		[m]
$c_i = 2.4300 \times 10^{-2}$	actuator value	[A]
$k_i = 2.5165$	actuator value	[A]
$x_{3MIN} = 3.8840 \times 10^{-2}$	limitation for current	[A]
$u_{MIN} = 4.9800 \times 10^{-3}$	limitation for voltage	

Variables x_{3MIN} and u_{MIN} are limitations for current and voltage to avoid break down of device. In this paper, the mathematical output has been compared simultaneously with NN and experimental results during real-time experiments. It is important to have appropriate conversion tables which include the experimental measurement to convert between quantities in order to have accuracy in the simulation.

3.1. Nn-Narx Modeling of the Ml System

Nature successfully builds up biological neural systems to perform control and adaptation tasks of living beings effectively in their complicated conditions and dynamically changing environments. To benefit from these assets, ANNs, which resemble biological neural systems, are widely utilized in intelligent system design due to the benefits of the learning and prediction potential of ANN algorithms.

Autoregressive neural network models can learn responses of linear and nonlinear dynamical systems from sampled input and output data [21,24]. Therefore, these models can gain a significant flexibility for multi-loop MRAC-FOPID control structures to be easily employed in real-world applications. NARX modeling can automatically learn response of a wide-range of system models in black box form and this reduces the need for guessing the model structure prior to model identification. The reference model is automatically identified from data that are captured online from the input and output of controlled systems. This point is an important contribution of this study to transform the multi-loop MRAC-FOPID control structures into an intelligent control system and thus increase effectiveness of MRAC-FOPID control in practical nonlinear control applications. For experimental validation, the proposed control structure was employed for the control of a magnetic levitation system. The magnetic levitation control introduces a highly nonlinear, unstable system control problem.

The authors observed that the performance of multi-loop MRAC-FOPID control with the MIT rule largely depends on the capability of the reference model to represent the leading closed-loop dynamics of the experimental ML system. In [9], a linear model of the experimental ML system was used to obtain the transfer function of the closed-loop retuning FOPID control system. This linear function was used as the reference model. This modeling approach comes out with two drawbacks for multi-loop MRAC-FOPID control structure which was already mentioned. In the current study, to deal with these drawbacks, NARX modeling was employed to obtain a more accurate reference model. The use of neural networks in system identification allows utilization of machine learning techniques in control, and this enables the multi-loop MRAC-FOPID control structure with NARX model to be more adaptive and intelligent compared to previous configurations.

Artificial neural networks can be successfully applied in the modeling of complex nonlinear systems, systems with disturbances and insufficiently known parameters, unpredictable and uncoordinated systems [19,56]. Figure 4 shows a block diagram of the NARX model. It is composed of time delay blocks (TDL) and conventional artificial neuron models with input and loop weight coefficients (IW and LW, respectively), bias (b), and activation function (f). The vector $u(t)$ is transmitted to the system as a two-layer input and is realized by passing a time delay block (b1) and in that way a part of Equation (33), is realized. The vector is obtained from the input weight coefficients defined in IW1,1, bias, b1, and the output network vector, $y(t)$, which leads to a time delay block and then multiplied by the weight coefficient, LW1,3. The resulting vector, $n1(t)$ is then directed to the first layer activation function, f1. The output from the first layer $a1(t)$, which represents the result of the operation of the activation function, f1, reaches the second layer, which is converted to the vector $n2(t)$ by the weight coefficients of LW2,1 and the b2 bias. This vector is then plotted in bias f2 to obtain the final neural network output, y(t). The feedback from the output neurons to the input of network causes a system state formation and lead to mimicing dynamical system models. Therefore, it can learn and represent time responses of dynamic systems. Two-layer neural networks with a nonlinear activation function are capable of learning nonlinear relations in inputs and state transition patterns. This property enables representation of nonlinear system dynamics by NARX model when it is trained properly. Therefore, after training with input-output data of the dynamical systems, NARX model can mimic dynamical behavior of systems and it is employed for prediction of time response of dynamic systems. This property makes NARX modeling a good candidate solution for online modeling problems of nonlinear dynamical systems in intelligent control applications.

The reason why a neural network is selected for modeling is that neural networks are very good at time-series problems. A neural network with enough elements (neurons) can model dynamic systems with arbitrary accuracy. They are particularly well suited for addressing non-linear dynamic problems. Neural network is a good candidate for solving this problem. The network will be designed by using recordings of an actual levitated magnet's position responding to a control current. The output of a NARX artificial neural network has the following form [20,23,24]:

$$y(t) = f\begin{pmatrix} y(t-1), y(t-2), ...y(t-n) \\ u(t-1), u(t-2), ...u(t-n) \end{pmatrix}, \tag{33}$$

where $y(t-i)$ is the ith previous sample of NARX output and $u(t-i)$ is the i-th previous sample of the NARX input. These samples are stored to the TDL buffers and the parameter n configures the buffer size that determines depth of delayed memory to be considered in the NARX model. The feedback from model output with previous values $[y(t-1), y(t-2), y(t-3),..,y(t-n)]$ forms an autoregressive model to predict the current value of the dynamical system. This delayed output values constitute pseudo-states to allow learning of system dynamics from time response data. The weight coefficients are indicated by lw, obtained vector is stated with n, first layer named $a_1(t)$, b is bias, and f is activation functions.

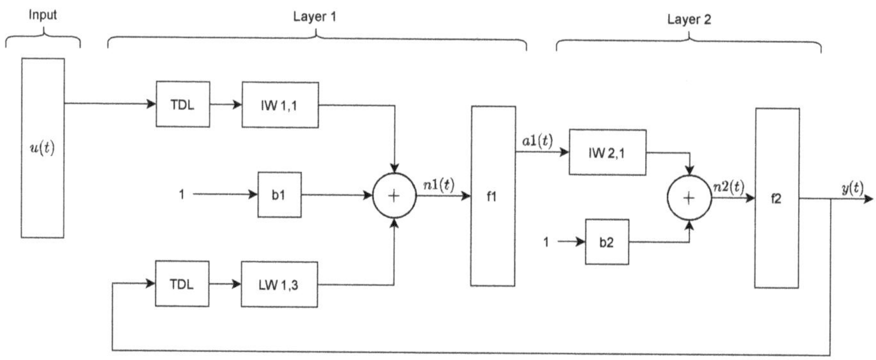

Figure 4. Representation of neural network (NN)-nonlinear autoregressive neural network with external inputs (NARX) structure. The first layer activation function f_1 is generally a nonlinear activation function to model nonlinear relations in the data. The second layer activation function f_2 is a linear activation function that rescales NARX output to original output data. Thus, it can yield satisfactory predictions of a nonlinear system dynamics.

To obtain a black-box model of the experimental ML system via NARX modeling, time-series data was collected from the output of controller and the sphere position sensors. The data from output of controller was used for input data ($u(t)$) of NARX model. A time-series signal for the sphere position was collected and used to form output data ($y(t)$) of NARX model. The data set $[u(t)y(t)]$ consists of 30,001 samples and it was used for training of the NARX model. The sampling period of data collection was 0.001 s and the data collection interval was continued for 30 s.

In this section, the real MLS system is presented in order to compare the simulated models. The real MLS model is located between two ADC cards as the hardware-in-the-loop part. Details of neural network is shown in Table 2. The NARX model used a one input and three output neural network with a input layer with 12 neurons and hidden layer with one neuron. The buffer size (n) set to 1, which provides a sample delaying of input and output data in TDL blocks. Figure 5 shows the input and output with filters. The reason for using the filter is that in the real model, the signal has noise and needs a filter. The same was done in the neural network model. In fact, the neural network model is the same as the maglev model with a raw signal without a filter.

After sending the simulated control signal to the hardware and acquiring the real signals (voltage) from the hardware, the signals are converted to corresponding position, [m] and current [A] by experimental look-up tables. Then, the signals are filtered to eliminate undesirable noises. The velocity of the ball is obtained by an inaccurate, but acceptable, memory element which applies a unity integration step delay. The output is the previous input value. In Figure 6, an overview of our project is shown. The mathematical model and ANN model were inserted in the real-time Simulink model in order to compare the three parts in the same situations. The input includes the range of sine wave, chirp wave with the frequency variation from 1 Hz to 6 Hz, the step input which contains all range of frequency and the constant input (steady state position is 0.011 m).

To determine this NARX configuration, several sizes of TDL and neuron numbers in the hidden layer were tested for data from the experimental system. The best modeling performance was obtained for 12 neurons in hidden layer and TDL units with one sample delay element.

The training algorithm used was the Levenberg–Marquardt method and the training process was completed in 113 iterations. The training task of the NARX model was carried out off-line by using the data set $[u(t)y(t)]$ that was collected from the experimental system. The time-series data set was automatically divided into training, validation and test sets. Figure 6 shows a Simulink model that was developed for the training of NARX model in a virtual closed-loop. For the closed-loop model identification, NARX model is put into a virtual control loop that is composed of the controller function of the identified closed-loop system and the NARX model to access the in-loop dynamics of the ML system. This virtual closed-loop can provide relevant training data, which is related to the in-loop dynamics of ML setup, to learn by NARX model. In the cases that outputs of the identified closed-loop experimental control system and the virtual closed-loop system with NARX model match, NARX model is said to be learned responses of ML system in the closed-loop. In the following section, this virtual closed-loop control system with NARX model is used as the reference model of multi-loop MRAC-FOPID control structure.

Table 2. The parameters of the neural network.

Parameter	Value
Sample rate	0.001 s
Simulation time	30 s
Hidden layer	12 neurons
X, input	Output of controller
Delay (external input for control current)	1
T, target	Output of system (ball position, velocity and coil current)
Iteration	1000
Reference signals	Pulse, Sine, Chirp (1 Hz to 6 Hz)

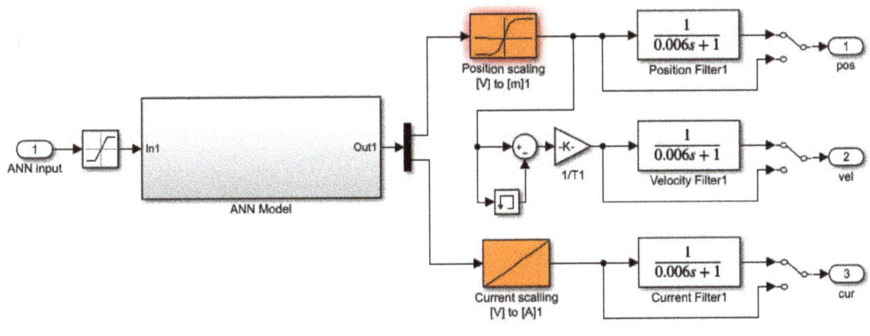

Figure 5. Neural network model with ML's filter.

All simulation results were obtained at Control Systems Research Laboratory, Tallinn University of Technology.

3.2. *Off-Line Results*

In this section, the trained ANN has been simulated. The output of the controller was saved and used for off-line simulation. Figure 7 represents the levitation characteristics (input control voltage and output position). Ball position of real system and ANN model are shown. The steady state position for real model is 0.01085 m. It is clearly shown that the ANN model tracks the input in steady state position, (0.0109 m) with about 0.5% error. However, in the transient regions there is considerable offset of about 7% error.

Figure 7 includes data when the ball is hit to create an external disturbance. The reason ANN response is seen in the present of disturbance is that the input of ANN (control output) is the same as real-model input. However, it is not realistic and the individual control output is used to control the ANN model in the real life experiment. The approximation of error in this noteworthy experiment gives mean square error (MSE) equal to $2.33 \cdot 10^{-7}$. The duration time of simulation is 30 s. The MSE is calculated as sum of difference between the samples of real ball position and ANN ball position dividing by number of samples (30,001 samples).

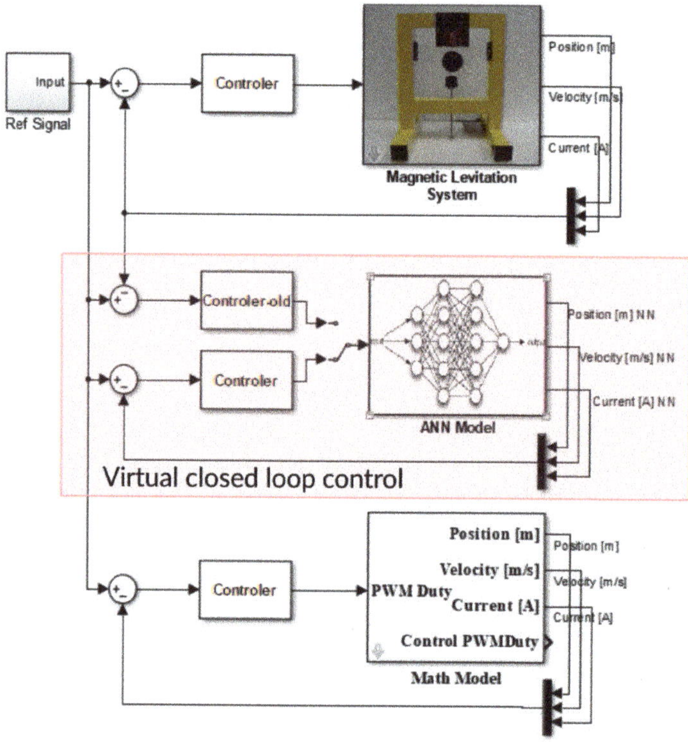

Figure 6. Overview of three parts, hardware, ANN model and Mathematical model.

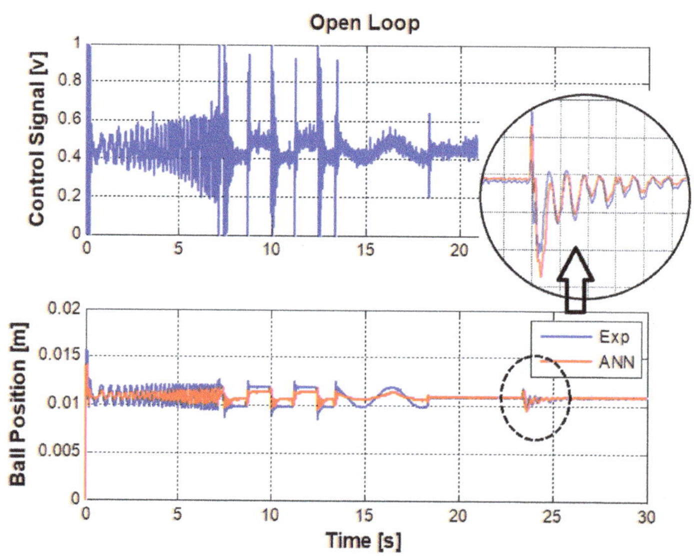

Figure 7. Input and output signals to train the NARX model for closed-loop model identification.

3.3. *Real-Life Results*

In this section, the whole system, including the mathematical model, ANN model, and real-time MLS, is simulated simultaneously. Since the system is fundamentally unstable, it means using the closed-loop system is essential. The input signals, including chirp, step, sinand constant source, are applied and used for three models. The control signal is generated by LQ controller and converted to PWM in real MLS.

The output of the ANN model is based on real-time controller. On the other hand, the controller already has been designed for real-time model. However, the obtained ANN is using the control signal that is being generated by real-model. The ANN model is not exactly like the real-model. It means a new model may need to implement a new controller in the simulation. Since the MLS is a highly nonlinear unstable system, it needs to illustrate complex control algorithms. The obtained ANN model can control itself by previous controller with minor changes if it works in linear region. The developed controller is applied to the system. The results are shown from Figures 8 and 9. In Figure 8, it is seen that the ball is exposed to disturbance in real life, that is why the simulation models can not respond to real life disturbance.

The figures show the higher frequency response of three models up to 7 s. It can be seen that the position response in real model is less than that of the mathematical and ANN models. In other words, the bandwidth of mathematical model is more than real-model. However, the ANN result indicates better behavior in high frequency.

In Figure 9, it can be seen that the transient response is compensated desirably and the steady state response of the ANN model is acceptable. However, for a high frequency region like chirp signal or step points in pulse signals performance has not been achieved well. The MSE error, in this case has been decreased considerably compared to the off-line ANN model.

In this paper, NN-NARX structure and linearization by dynamic output feedback were implemented to control ball position of a real magnet levitation system. The noisy signals (signals before filters) are used in order to train the neural network model more accurately.

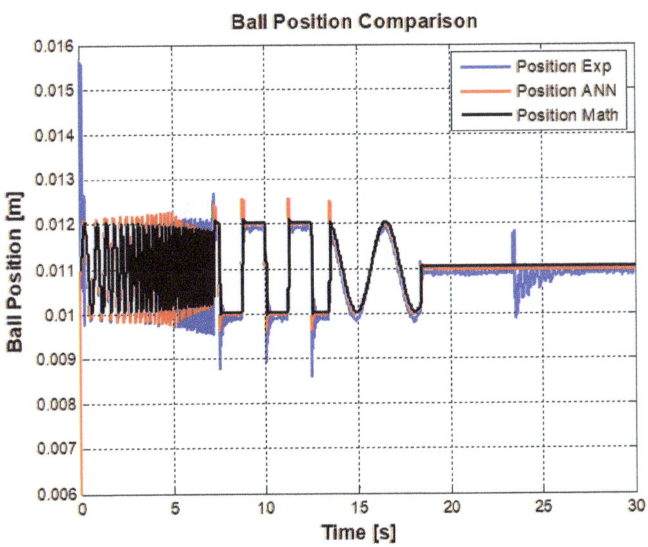

Figure 8. Ball position comparison among real model, ANN model and Mathematical model.

The Levenberg–Marquardt method gives the best results with 12 neurons in the hidden layer and one delay. The training process achieved best performances after 113 iterations. Increasing the number of neurons in the hidden layer in this project did not improve the accuracy of the model. As the input signal in closed-loop system depends on the error between input and its output, the ANN controller is developed instead of real MLS LQ controller. Results indicate the steady state and transient response is compensated desirably for the neural network model.

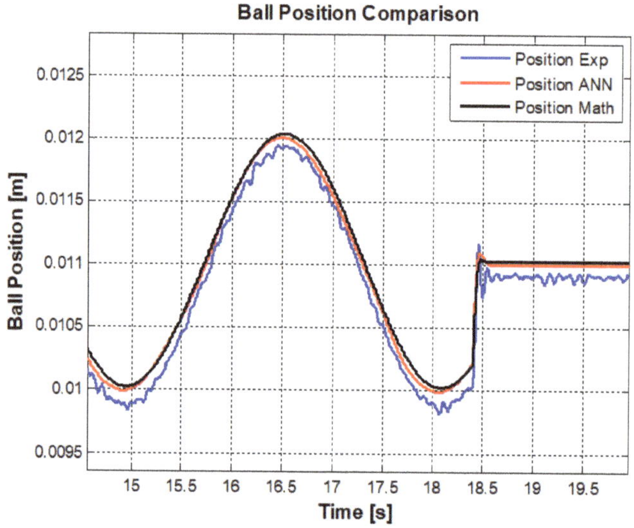

Figure 9. A comparison of ANN model performance and mathematical model performance.

For the purpose of comparing all models, the mean square error between the input command and output of models was calculated. Table 3 shows the results where YEXP, YANN, and YMATH represent the ball position in the experiment, artificial neural network, and mathematical model, respectively. UEXP is the input command in the experimental test. Based on results in the figures cited above and Table 3, it is concluded that the identified model has acceptable performance.

Table 3. MSE ball position comparison among models.

Compare Type	MSE
ANN Model off-line (YEXP-YANN)	2.32×10^{-7}
Real-Time Simulation (YEXP-UEXP)	2.12×10^{-7}
Mathematical Model (YMATH-UEXP)	8.02×10^{-8}
ANN Model (YANN-UEXP)	2.83×10^{-7}

4. Multi-Loop Mrac-Fopid Control with Narx Reference Model for Ml System Control Application

This section presents the experimental results of multi-loop MRAC-FOPID control of ML system by using a NARX reference model. The design steps of Multi-loop MRAC-FOPID control was carried out for experimental ML system:

i. The closed-loop retuning FOPID control system was implemented and FOPID controller was optimally tuned by using FOMCON toolbox [47].
ii. The control signal and sphere position data from the designed closed-loop control system described above were collected and these data were used to train the NARX model in the virtual closed-loop control system. Thus, the virtual closed-loop PID control loop with the NARX model is used as reference model to represent closed-loop retuning FOPID control system.
iii. The outer loop is connected to inner loop according to MIT rule as shown in Figure 10.

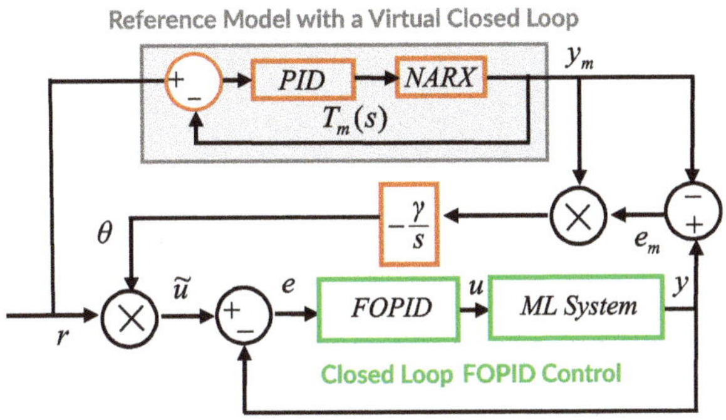

Figure 10. Block diagram of multi-loop MRAC-FOPID control with a NARX reference model.

Regarding the configuration of the experimental system, in Figure 11, the complete control diagram used to implement both simulation and real time experiments is presented. For the implementation of the experimental configuration, the MATLAB/Simulink environment is used with FOMCON toolbox for implementing fractional-order control. The tuning of the controllers used in the retuning loop is detailed in [9]. However, compared to that work, the value of γ in the MRAC loop was tuned down from $-40,000$ to $-20,000$ in order to minimize high frequency components entering the control system and causing excessive vibrations of the sphere.

Figure 12 shows a simulation result to demonstrate disturbance rejection performance of multi-loop MRAC-FOPID with NARX model. After settling of the sphere to a height of 0.0109 m, a step-down disturbance and then step-up input disturbances were applied at the simulation times 10 s and 40 s, respectively. The responses of the proposed multi-loop MRAC-FOPID and the conventional FOPID control are shown in the figure. The conventional FOPID control is the inner loop that was performed by disabling contribution of outer loop (MRAC with MIT rule). Thus we can test control improvements of the multi-loop MRAC-FOPID with the NARX model compared to the FOPID control loop alone. At disturbance incidents, multi-loop MRAC-FOPID with NARX model settles back to the set-point faster with lower ripples. These results indicate considerable improvements indisturbance rejection control performance of the FOPID control loop when the outer loop (MRAC) is enabled.

Armed with promising simulation results, real-time verification of the multi-loop control system was performed. The Simulink model was built according to Figure 11 with additive disturbance injected into the system using the control input $u(t)$ in the original PID control loop. The model runs on a Desktop PC (3.40 GHz Processor and 8.00 GB RAM) and the computational complexity of the control scheme allows real-time operation with 0.001 data sampling rate. Figures 13 and 14 show experimental results to demonstrate disturbance rejection performance of the multi-loop MRAC-FOPID control with NARX reference in comparison with the conventional FOPID control (MRAC is disabled). In Figure 13, after settling to the set-point of 0.0109 m high, a step-down input disturbance signal was applied at 10 s. Responses of control systems to this disturbance signal are compared in the figure. The multi-loop MRAC-FOPID control improves disturbance responses of ML system with faster resettling and lower ripples around the set-point. Figure 14 shows impacts of a continuous input disturbance in the form of sinusoidal waveform on the set-point control performance. The sinusoidal disturbance with roughly 7 s periods was applied. The experimental results show that the multi-loop MRAC-FOPID control with NARX reference modeling better suppresses the sinusoidal disturbance at the system output compared the FOPID control loop. These two tests evidently verify contribution of outer loop (adaptation loop) to disturbance rejection performance of the inner loop (control loop). Results also validate the improvement of disturbance rejection control capacity without any deterioration of set-point control performance. The results are also summarized in Table 4. This property is a natural result of designing loops with specialized objectives in multi-loop structure. This control structure can be a feasible solution of the tradeoff between set-point and disturbance rejection performances of single loop systems.

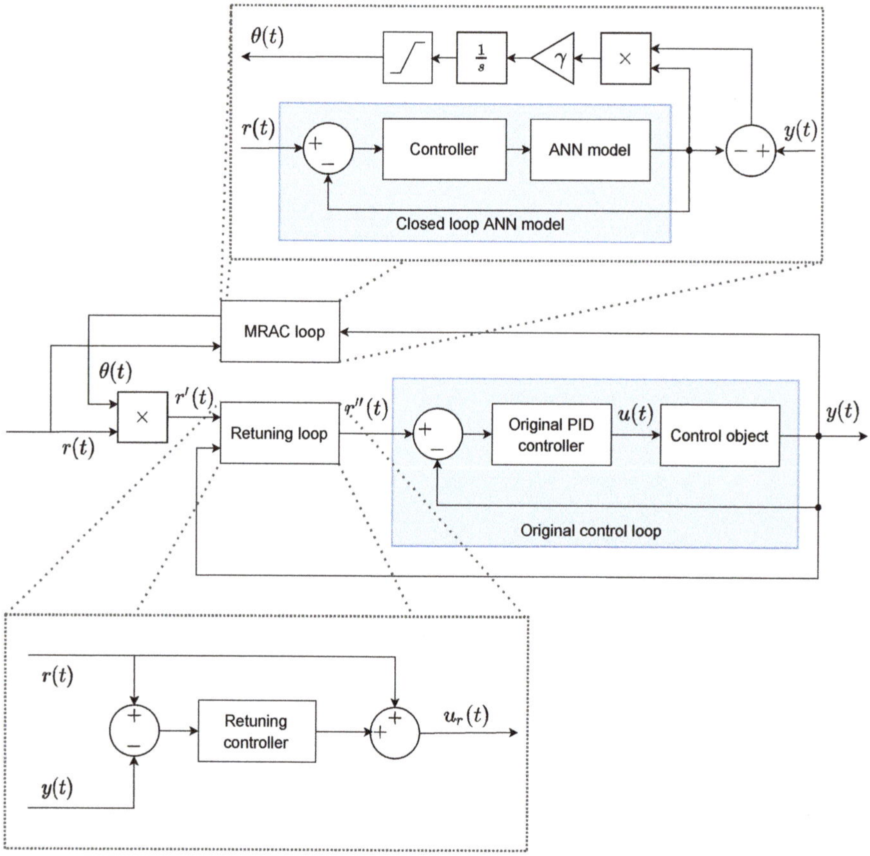

Figure 11. Complete experimental configuration for evaluating the multi-loop control structure. There are three control loops in total: the original PID control loop with reference input $r''(t)$, the retuning loop with reference input $r'(t)$ that replaces the dynamics of the original loop with those of the optimally tuned FOPID controller, and the MRAC loop to which the original reference input $r(t)$ is connected. By bypassing reference inputs in various ways, it is possible to achieve different simulation scenarios with the MRAC loop and retuning control loop enabled or disabled independently. This schematic diagram serves as the basis for both pure software simulations and real-time experiments in MATLAB/Simulink software.

In addition, experiments involving the original PID control loop with MRAC adaptation loop were performed in the same way, i.e., for two different types of disturbance and with MRAC loop enabled and disabled. The results of the successful experiments are presented in Figures 15 and 16 and also in Table 4. Only those experiments that had the MRAC loop enabled were successful, so only those are presented. In other cases, the control loop failed to levitate the sphere and keep it in a steady state—in the real experiment the sphere was simply thrown out from the working area due to unstable behavior of the control system. The original PID control loop only works in the case when the sphere is manually placed in the magnetic field away from the rest, whereas other types of control configurations are also able to successfully levitate the sphere from its initial position on the rest. Taking the results into account, one can see that even with the MRAC loop enabled, the original PID control loop although capable of levitating the sphere, still shows inferior performance compared to the fractional retuning control with the MRAC loop enabled.

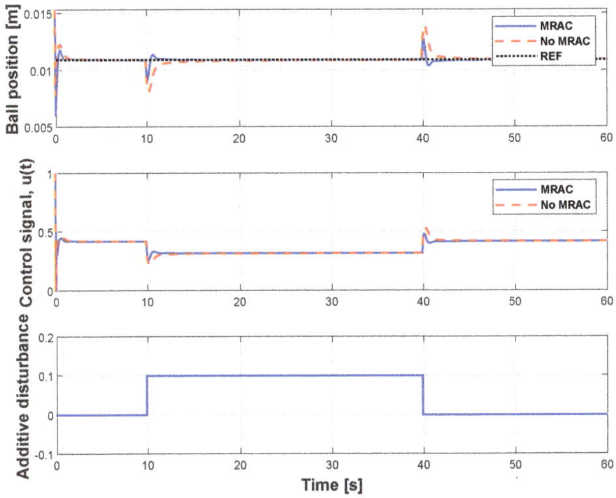

Figure 12. Disturbance responses of the multi-loop MRAC-FOPID control with NARX reference model and the FOPID control loop (When MRAC is disabled).

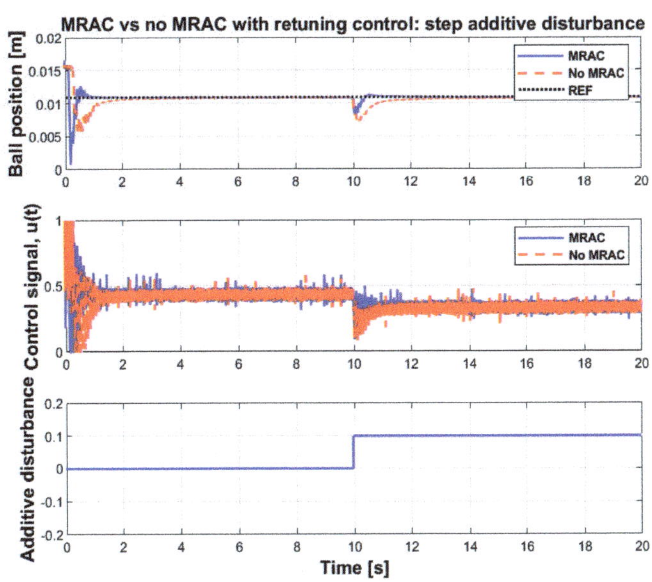

Figure 13. Step disturbance responses of the multi-loop MRAC-FOPID control with NARX reference model and the FOPID control loop (When MRAC is disabled).

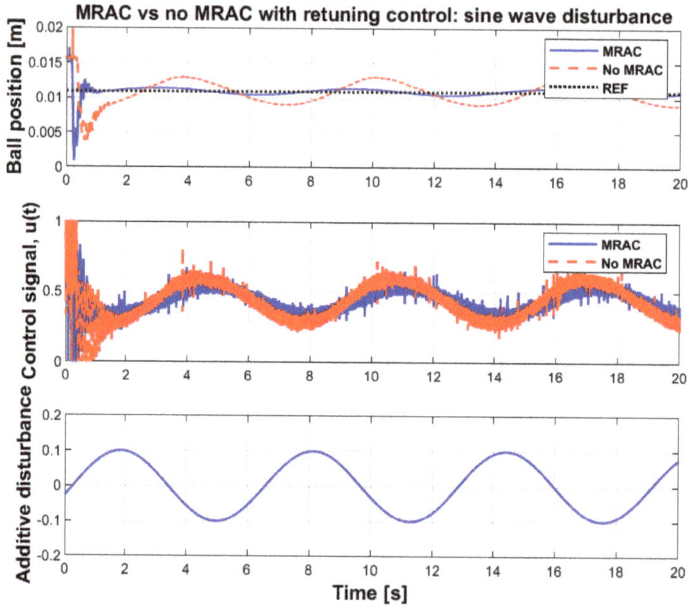

Figure 14. Sinusoidal disturbance responses of the multi-loop MRAC-FOPID control with NARX reference model and the FOPID control loop (When MRAC is disabled).

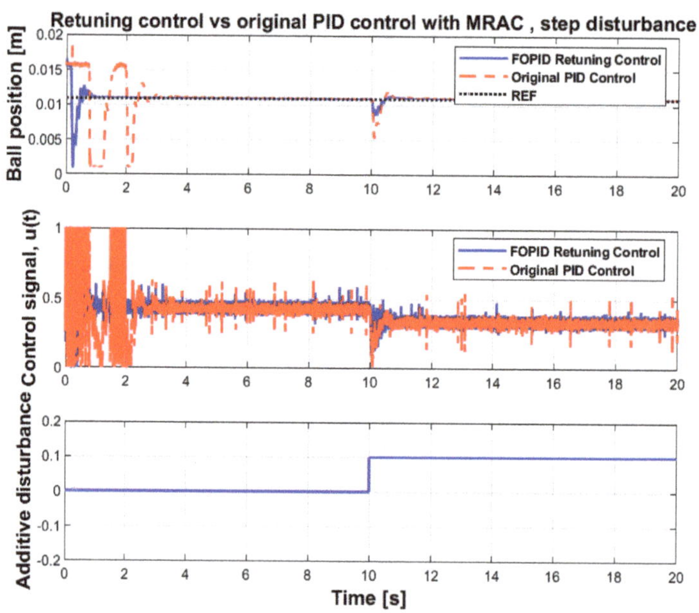

Figure 15. Step disturbance responses of the MRAC system with the original PID control loop.

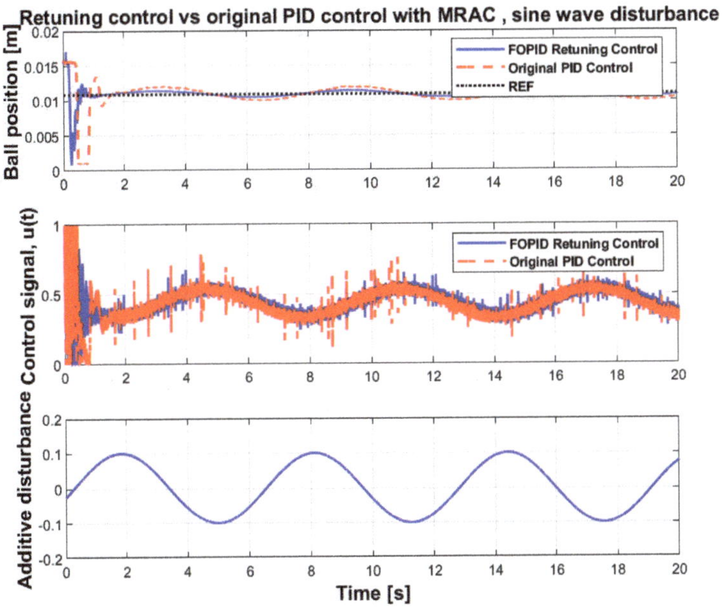

Figure 16. Sine wave disturbance responses of MRAC system with original PID control loop.

Table 4. Comparison of experimental performances of control systems with the MRAC loop disabled and enabled.

Control Structure	Peak Values of$\lvert y(t) - r(t) \rvert$ in Disturbance Responses [m]	Settling Time after Step Disturbance [s]	Cumulative Absolute Control Error (Additive Disturbance)	Cumulative Absolute Control Error (Harmonic Disturbance)
FOPID control loop (MRAC disabled)	3.603×10^{-3}	1.23	8.6403	28.6592
Multi-loop MRAC-FOPID control with NARX reference model	2.021×10^{-3}	0.43	3.6240	8.2300
Multi-loop original PID control with NARX reference model	5.552×10^{-3}	1.08	16.3261	18.4087

5. Conclusions and Discussion

This experimental study investigated the use of multi-loop MRAC-FOPID control with NARX reference modeling in the disturbance rejection control problem of ML systems. Simulation and experimental studies were conducted and it was observed that NARX reference modeling enables more intelligent realization of multi-loop MRAC-FOPID control structures. Simulation and experimental studies reveal that NARX modeling can be utilized for online modeling of nonlinear, unstable dynamical systems and it can be a feasible solution for real-time modeling requirements of intelligent control systems. This study also shows that the recurrent artificial neural network serves for improvement of classical control loops.

It is remarkable that in this study, three major components were considered, which can be used independently to formulate a robust and well-performing control system: (1) the NARX-based control loop with a PID controller that was tuned separately from the FOPID control in the main loop that can be used as an accurate reference model of the controlled process; (2) the retuning FOPID control which replaces the dynamics of another existing PID controller; (3) the MRAC loop which together with the reference model results in improved disturbance rejection in the overall control loop.

Indeed, the resulting multi-loop MRAC-FOPID control with NARX reference modeling combines robust stability and set-point control performance of the FOPID control loop with adaptation skills of MRAC loop. Disturbance rejection control is a major concern for practical control applications. To maintain optimal control performance in real-world application, disturbance rejection control performance should be improved in addition to set-point and stability performance. The multi-objective control structures with specialized loops can provide more robust control performance than single control loops in real-world applications. Moreover, this multi-loop control structure can be easily applied to existing closed-loop control systems and this can give an opportunity to upgrade existing classic control loops to multi-loop MRAC loops without changing any parameter or any block of the existing control loops. The NARX reference model facilitates this upgrade process by performing real-time closed-loop model identification of the system and makes the multi-loop MRAC-FOPID control structure more adaptive and intelligent by providing it with benefits stemming from contemporary machine learning.

The MIT rule presents a shortcoming of being sensitive to output amplitude of the reference input [5,11]. Very high amplitudes may cause instability of the system [5] and very low amplitudes diminish effectiveness of MIT rule to respond disturbances [11]. In [11], several alternative multi-loop control structures have been suggested to deal with these drawbacks. Future work can be planned for practical realization and performance evaluations of multi-loop control structure with NARX reference modeling.

Author Contributions: conceptualization, A.T., E.P., K.V. and H.A.; methodology A.T., E.P., K.V, and B.B.A.; validation, H.A. and A.T.; writing—original draft preparation, A.T., B.B.A. and H.A.; writing—review and editing, E.P. and K.V; supervision, A.T. and E.P. All authors have read and agreed to the published version of the manuscript.

Funding: The work was supported by the Estonian Research Council grant PRG658.

Conflicts of Interest: The authors declare no conflict of interest. The funders had no role in the design of the study; in the collection, analyses, or interpretation of data; in the writing of the manuscript, or in the decision to publish the results.

References

1. Alagoz, B.B.; Deniz, F.N.; Keles, C.; Tan, N. Implicit disturbance rejection performance analysis of closed loop control systems according to communication channel limitations. *IET Control Theory Appl.* **2015**, *9*, 2522–2531. [CrossRef]
2. Butler, H. In Proceedings of the *Model Reference Adaptive Control: From Theory to Practice*; Prentice Hall International Series in Systems and Control Engineering; Prentice Hall: Upper Saddle River, NJ, USA, 1992.
3. Vinagre, B.; Petráš, I.; Podlubny, I.; Chen, Y. Using fractional order adjustment rules and fractional order reference models in model-reference adaptive control. *Nonlinear Dyn.* **2002**, *29*, 269–279. [CrossRef]
4. Lavretsky, E. Adaptive control: Introduction, overview, and applications. In Proceedings of the NASA Adaptive Control Workshop, NASA Marshall Space Center, Huntsville, AL, USA, 24 February 2009.
5. Jain, P.; Nigam, M. Design of a Model Reference Adaptive Controller Using Modified MIT Rule for a Second Order System. *Adv. Electron. Electr. Eng.* **2013**, *3*, 477–484.
6. Swamkar, P.; Nema, R.; Swarnkar, P.; Kumar, J.; Nema, R. Comparative Analysis of MIT Rule and Lyapunov Rule in Model Reference Adaptive Control Scheme. *Innov. Syst. Des. Eng.* **2011**, *2*, 154–162.
7. Kavuran, G.; Ates, A.; Alagoz, B.B.; Yeroglu, C. An experimental study on model reference adaptive control of TRMS by error-modified fractional order MIT rule. *Control Eng. Appl. Inform.* **2017**, *19*, 101–111.

8. Alagoz, B.B.; Tepljakov, A.; Petlenkov, E.; Yeroglu, C. Multi-Loop model reference adaptive control of fractional-order PID control systems. In Proceedings of the 2017 IEEE 40th International Conference on Telecommunications and Signal Processing (TSP), Barcelona, Spain, 5–7 July 2017. [CrossRef]
9. Tepljakov, A.; Alagoz, B.B.; Gonzalez, E.; Petlenkov, E.; Yeroglu, C. Model Reference Adaptive Control Scheme for Retuning Method-Based Fractional-Order PID Control with Disturbance Rejection Applied to Closed-Loop Control of a Magnetic Levitation System. *J. Circuits Syst. Comput.* **2018**, *27*. [CrossRef]
10. Rajesh, R.; Deepa, S. Design of direct MRAC augmented with 2 DoF PIDD controller: An application to speed control of a servo plant. *J. King Saud Univ.-Eng. Sci.* **2020**, *32*, 310–320. [CrossRef]
11. Alagoz, B.B.; Tepljakov, A.; Petlenkov, E.; Yeroglu, C. Multi-Loop Model Reference Proportional Integral Derivative Controls: Design and Performance Evaluations. *Algorithms* **2020**, *13*, 38. [CrossRef]
12. Chen, Y.; Petras, I.; Xue, D. Fractional order control-a tutorial. In Proceedings of the 2009 IEEE American Control Conference, St. Louis, MO, USA, 10–12 June 2009. [CrossRef]
13. Monje, C.A.; Chen, Y.; Vinagre, B.M.; Xue, D.; Feliu, V. *Fractional-Order Systems and Controls*; Springer: London, UK, 2010; doi:10.1007/978-1-84996-335-0. [CrossRef]
14. Podlubny, I. *Fractional Differential Equations*; Mathematics in Science and Engineering; Academic Press: San Diego, CA, USA, 1999.
15. Zhao, C.; Xue, D.; Chen, Y.Q. A fractional order PID tuning algorithm for a class of fractional order plants. In Proceedings of the IEEE International Conference Mechatronics and Automation, Niagara Falls, ON, Canada, 29 July–1 August 2005; Volume 1, pp. 216–221. [CrossRef]
16. Padula, F.; Visioli, A. *Advances in Robust Fractional Control*; Springer: Cham, Switzerland, 2014.
17. Tepljakov, A.; Petlenkov, E.; Belikov, J.; Gonzalez, E.A. Design of retuning fractional PID controllers for a closed-loop magnetic levitation control system. In Proceedings of the 2014 13th International Conference on Control, Automation, Robotics & Vision, ICARCV, Singapore, 10–12 December 2014; pp 1345–1350.
18. Tepljakov, A.; Gonzalez, E.A.; Petlenkov, E.; Belikov, J.; Monje, C.A.; Petráš, I. Incorporation of fractional-order dynamics into an existing PI/PID DC motor control loop. *ISA Trans.* **2016**, *60*, 262–273. [CrossRef]
19. Trisanto, A. PID controllers using neural network for multi-input multioutput magnetic levitation system. *Electrician* **2007**, *1*, 62–68.
20. Vassiljeva, K.; Tepljakov, A.; Petlenkov, E. NN-ANARX model based control of liquid level using visual feedback. In Proceedings of the 2015 IEEE 13th International Conference on Industrial Informatics (INDIN), Cambridge, UK, 22–24 July 2015. [CrossRef]
21. Jiang, Y.; Yang, N.; Yao, Q.; Wu, Z.; Jin, W. Real-time moisture control in sintering process using offline–online NARX neural networks. *Neurocomputing* **2020**, *396*, 209–215. [CrossRef]
22. Abouelazayem, S.; Glavinić, I.; Wondrak, T.; Hlava, J. Adaptive Control of Meniscus Velocity in Continuous Caster based on NARX Neural Network Model. *IFAC-PapersOnLine* **2019**, *52*, 222–227. [CrossRef]
23. Petlenkov, E.; Nomm, S.; Kotta, U. Neural Networks Based ANARX Structure for Identification and Model Based Control. In Proceedings of the 2006 IEEE 9th International Conference on Control, Automation, Robotics and Vision, Singapore, 5–8 December 2006. [CrossRef]
24. Antić, D.; Milovanović, M.; Nikolić, S.; Milojković, M.; Perić, S. Simulation Model of Magnetic Levitation Based on NARX Neural Networks. *Int. J. Intell. Syst. Appl.* **2013**, *5*, 25–32. [CrossRef]
25. Kumpati, S.N.; Kannan, P.; others. Identification and control of dynamical systems using neural networks. *IEEE Trans. Neural Netw.* **1990**, *1*, 4–27.
26. Hagan, M.T.; Demuth, H.B. Neural Networks for Control. In Proceedings of the 1999 American Control Conference, Hyatt Regency San Diego, San Diego, CA, USA, 2–4 June 1999; pp. 1642–1656.
27. Slotine, J.J.E.; Li, W.; others. *Applied Nonlinear Control*; Prentice Hall: Englewood Cliffs, NJ, USA, 1991; Volume 199.
28. Spooner, J.T.; Maggiore, M.; Ordonez, R.; Passino, K.M. *Stable Adaptive Control and Estimation for Nonlinear Systems: Neural and Fuzzy Approximator Techniques*; John Wiley & Sons: Hoboken, NJ, USA, 2004; Volume 43.
29. Fatemimoghadam, A.; Toshani, H.; Manthouri, M. Control of magnetic levitation system using recurrent neural network-based adaptive optimal backstepping strategy. *Trans. Inst. Meas. Control.* **2020**, 0142331220911821. [CrossRef]

30. Hajimani, M.; Gholami, M.; Dashti, Z.A.S.; Jafari, M.; Shoorehdeli, M.A. Neural adaptive controller for magnetic levitation system. In Proceedings of the 2014 IEEE Iranian Conference on Intelligent Systems (ICIS), Bam, Iran, 4–6 February 2014; pp. 1–6.
31. Bidikli, B. An observer-based adaptive control design for the maglev system. *Trans. Inst. Meas. Control.* **2020**, 0142331220932396. [CrossRef]
32. Zhang, J.; Wang, X.; Shao, X. Design and real-time implementation of Takagi–Sugeno fuzzy controller for magnetic levitation ball system. *IEEE Access* **2020**, *8*, 38221–38228. [CrossRef]
33. Ladaci, S.; Charef, A. On fractional adaptive control. *Nonlinear Dyn.* **2006**, *43*, 365–378. [CrossRef]
34. Wei, Y.; Sun, Z.; Hu, Y.; Wang, Y. On fractional order composite model reference adaptive control. *Int. J. Syst. Sci.* **2016**, *47*, 2521–2531. [CrossRef]
35. Aguila-Camacho, N.; Duarte-Mermoud, M.A. Fractional adaptive control for an automatic voltage regulator. *ISA Trans.* **2013**, *52*, 807–815. [CrossRef]
36. Keziz, B.; Ladaci, S.; Djouambi, A. Design of a MRAC-Based Fractional order PI λ D μ Regulator for DC Motor Speed Control, In Proceedings of the 3rd International Conference on Electrical Sciences and Technologies in Maghreb (CISTEM 2018), Algiers, Algeria, 28–31 October 2018; pp. 1–6.
37. Hartley, T.T.; Lorenzo, C.F. Dynamics and Control of Initialized Fractional-Order Systems. *Nonlinear Dyn.* **2002**, *29*, 201–233.:1016534921583. [CrossRef]
38. Vinagre, B.M.; Podlubny, I.; Hernández, A.; Feliu, V. Some Approximations Of Fractional Order Operators Used In Control Theory And Applications. *Fract. Calc. Appl. Anal.* **2000**, *3*, 945–950.
39. Matignon, D. Stability Results For Fractional Differential Equations With Applications to Control Processing. *Comput. Eng. Syst. Appl.* **1997**, *2*, 963–968.
40. Radwan, A.; Soliman, A.; Elwakil, A.; Sedeek, A. On the stability of linear systems with fractional-order elements. *Chaos Solitons Fractals* **2009**, *40*, 2317–2328. [CrossRef]
41. Senol, B.; Ates, A.; Baykant Alagoz, B.; Yeroglu, C. A numerical investigation for robust stability of fractional-order uncertain systems. *ISA Trans.* **2014**, *53*, 189–198. [CrossRef] [PubMed]
42. Alagoz, B.B. Hurwitz stability analysis of fractional order LTI systems according to principal characteristic equations. *ISA Trans.* **2017**, *70*, 7–15. [CrossRef]
43. Chen, Y.; Vinagre, B.M.; Podlubny, I. Continued Fraction Expansion Approaches to Discretizing Fractional Order Derivatives?an Expository Review. *Nonlinear Dyn.* **2004**, *38*, 155–170. [CrossRef]
44. Oustaloup, A.; Levron, F.; Mathieu, B.; Nanot, F. Frequency-band complex noninteger differentiator: Characterization and synthesis. *IEEE Trans. Circuits Syst. Fundam. Theory Appl.* **2000**, *47*, 25–39. [CrossRef]
45. Deniz, F.N.; Alagoz, B.B.; Tan, N.; Koseoglu, M. Revisiting four approximation methods for fractional order transfer function implementations: Stability preservation, time and frequency response matching analyses. *Annu. Rev. Control.* **2020**, *49*, 239–257. [CrossRef]
46. Tepljakov, A. Implementation of Fractional-Order Models and Controllers. In *Fractional-order Modeling and Control of Dynamic Systems*; Springer International Publishing: Berlin/Heidelberg, Germany, 2017; pp. 77–105. [CrossRef]
47. Tepljakov, A.; Petlenkov, E.; Belikov, J. FOMCON toolbox for modeling, design and implementation of fractional-order control systems. In *Applications in Control*; Petráš, I., Ed.; De Gruyter: Berlin, Germany, 2019; pp. 211–236. [CrossRef]
48. bo Zhou, H.; an Duan, J. Levitation mechanism modelling for maglev transportation system. *J. Cent. South Univ. Technol.* **2010**, *17*, 1230–1237. [CrossRef]
49. Yan, L. Development and Application of the Maglev Transportation System. *IEEE Trans. Appl. Supercond.* **2008**, *18*, 92–99. [CrossRef]
50. Dragos, C.A.; Preitl, S.; Precup, R.E.; Bulzan, R.G.; Petriu, E.M.; Tar, J.K. Experiments in fuzzy control of a Magnetic Levitation System laboratory equipment. In Proceedings of the IEEE 8th International Symposium on Intelligent Systems and Informatics, Subotica, Serbia, 10–11 September 2010. [CrossRef]
51. Gole, H.; Barve, P.; Kesarkar, A.A.; Selvaganesan, N. Investigation of fractional control performance for magnetic levitation experimental set-up. In Proceedings of the 2012 IEEE International Conference on Emerging Trends in Science, Engineering and Technology (INCOSET), Tiruchirappalli, India, 13–14 December 2012. [CrossRef]
52. Yu, P.; Li, J.; Li, J. The Active Fractional Order Control for Maglev Suspension System. *Math. Probl. Eng.* **2015**, *2015*, 1–8. [CrossRef]

53. Muresan, C.I.; Ionescu, C.; Folea, S.; Keyser, R.D. Fractional order control of unstable processes: The magnetic levitation study case. *Nonlinear Dyn.* **2014**, *80*, 1761–1772. [CrossRef]
54. Dolga, V.; Dolga, L. Modeling and simulation of a magnetic levitation system. *Ann. Oradea. Univ. Fascicle Manag. Technol. Eng.* **2007**, *6*, 16.
55. Inteco Sp. z o.o. INTECO Official Webpage. Available online: http://www.inteco.com.pl/ (accessed on 13 July 2015).
56. Rafiq, M.Y.; Bugmann, G.; Easterbrook, D.J. Neural network design for engineering applications. *Comput. Struct.* **2001**, *79*, 1541–1552. [CrossRef]

algorithms

MDPI

Article

PD Steering Controller Utilizing the Predicted Position on Track for Autonomous Vehicles Driven on Slippery Roads

Natalia Alekseeva *, Ivan Tanev and Katsunori Shimohara

Graduate School of Science and Engineering, Doshisha University, Kyoto 602-8580, Japan; itanev@mail.doshisha.ac.jp (I.T.); kshimoha@sil.doshisha.ac.jp (K.S.)
* Correspondence: alexeevanatalia8@gmail.com

Received: 24 December 2019; Accepted: 18 February 2020; Published: 21 February 2020

Abstract: Among the most important characteristics of autonomous vehicles are the safety and robustness in various traffic situations and road conditions. In this paper, we focus on the development and analysis of the extended version of the canonical proportional-derivative PD controllers that are known to provide a good quality of steering on non-slippery (dry) roads. However, on slippery roads, due to the poor yaw controllability of the vehicle (suffering from understeering and oversteering), the quality of control of such controllers deteriorates. The proposed predicted PD controller (PPD controller) overcomes the main drawback of PD controllers, namely, the reactiveness of their steering behavior. The latter implies that steering output is a direct result of the currently perceived lateral- and angular deviation of the vehicle from its intended, ideal trajectory, which is the center of the lane. This reactiveness, combined with the tardiness of the yaw control of the vehicle on slippery roads, results in a significant lag in the control loop that could not be compensated completely by the predictive (derivative) component of these controllers. In our approach, keeping the controller efforts at the same level as in PD controllers by avoiding (i) complex computations and (ii) adding additional variables, the PPD controller shows better quality of steering than that of the evolved (via genetic programming) models.

Keywords: autonomous vehicles; automated steering; slippery road conditions; PD controllers; predictive model

1. Introduction

Essentially every year, the demand for autonomously controlled road motor vehicles (hereafter referred to as cars) is rising, and now they could be used both as a taxi [1] or as personal cars. Consequently, the demand for precise control models that provide the safest and fastest transit of the passengers to their destinations is growing. Hereinafter, we consider a control model to be a control feedback mechanism, the description of which we will provide in the following sections. Among the main aspects of such models, the automated control of steering of the car is achieved by continuously adjusting the steering angle of the front wheels of the car. At the moment, the PD controllers are among the most widely used for the steering control of autonomous cars [2,3]. Despite being generally effective under the ordinary dry road conditions and simple to implement, these controllers suffer from several drawbacks. One of them is that, due to the simplicity of structure and low number of variables, they cannot properly cooperate with the physics of the vehicle in the case of a slippery road [2]. When humans drive the car, we dynamically adapt our steering behavior depending on the features of the car (e.g., length, width, mass,) and the road conditions (dry, wet, snowy, etc.) in a way that is difficult to mimic in both PD and PID controllers due to their hard structure with a small number of variables. In addition, the reactivity of these controllers implies that the steering output is a direct result of the currently perceived lateral- and angular deviation of the car from its intended, ideal trajectory. For convenience, here, as in previous studies, we consider the middle of the lane to be the desired ideal trajectory. Because these deviations are used as an error in the error-correcting, feedback control of the steering of the car, the required non-zero value of the error during cornering would result in a trajectory of the car that is always offset to the "outside" of the corner. Consequently, if the turning is initiated as an obstacle-avoiding maneuver, the car will inevitably circumnavigate the obstacle at a distance that is always lower than that of the intended, ideal trajectory, which, in turn, leads to an increased risk of a collision with the obstacle. Moreover, the reactiveness, combined with the tardiness of the yaw control of the car on slippery roads (as a result of the significant reduction of the steering forces—due to the reduced friction coefficient between the tires and the road—that have to overcome the given non-zero yaw moment of inertia of the car) results in a significant lag in the control loop that could not be compensated completely by the predictive (derivative) component of these controllers.

Another challenge of adopting the PD and PID controllers is finding the optimal values of the scaling coefficients of these controllers for the particular road conditions. The optimization of these parameters by human experts often requires extensive knowledge in both the control theory and vehicle dynamics. Automated tuning of the parameters, on the other hand, might require applying heuristic approaches that are notorious with their long runtime even when using a significant computational power [4,5].

In order to address the above-mentioned challenges of the canonical PD controllers, in which the control output is calculated as a weighted sum of the control errors and their derivatives, in our previous research we proposed a PID steering controller featuring an arbitrary (rather than an additive) internal structure, developed heuristically via genetic programming [6].

Another approach to improve the PD controller is adding prediction mechanics. Predictive models were already widely used in application to autonomous vehicles [7] for non-slippery road conditions. One of the most widely used methods is the Model Predictive Control (MPC) [8] and its modifications [9,10]. This method, however, features some drawbacks, which would hinder its applicability to the considered application. One of them is the computational overhead associated with (i) the need to predict too far ahead and (ii) the significant complexity of the predictive model (which would be even greater in slippery road conditions due to the complex nature of the vehicle dynamics of the sliding car).

In this work, we modified the original PD controller by replacing one of the terms with its predicted value. We tried to compensate for the drawbacks of the controller by avoiding (i) adding additional variables and (ii) modifying the structure of the controller that would increase the controller effort. Rather, by using the predicted (instead of the current) value of just one perceived variable, pertinent to the state of the car—the lateral deviation from the center of the road—we demonstrated that the quality of steering of the car on slippery roads could be significantly improved with the same set of perception information of the controller; yet, assuming the availability of the map of the road ahead.

The remainder of this paper is organized as follows. Section 2 explains the materials and methods of our research. In Section 3, we present the experimental results. Section 4 discusses the experimental results, and Section 5 draws a conclusion.

2. Materials and Methods

2.1. Environment and the Car Simulator TORCS

In this work, TORCS [7] was employed to perform a simulation of the experiments. This tool provides an accurate simulation of both the physics environment and mechanics of a car (engine, etc.). For the experiments, a racing model of a rear-wheel-drive car of the Mercedes brand was used, and a list of its parameters is shown in Table 1 below, and Figure 1 shows its view during the races.

Table 1. Main parameters of the simulated car.

Feature	Value
Model	CLK DTM
Length, m	4.76
Width, m	1.96
Height, m	1.17
Mass, kg	1050
Front/rear weight repartition	0.5/0.5
Height of center of gravity, m	0.25
Coefficient of friction of tires	1.0
Drivetrain	Front engine, rear wheels drive

Figure 1. Snapshots of the simulated car.

As we mentioned earlier [6], the choice of TORCS over different alternatives as a simulator in our experiments was also determined by its computational efficiency, safety, and the availability of its source code.

2.2. The Track

In our experiments, a route called a "fish hook" was constructed (Figure 2). Its length is only 300 m, but its shape contains a straight line, a sharp transition to the left turn, and a long right turn afterward. Such a track belongs to a difficult type of tracks for a human driver.

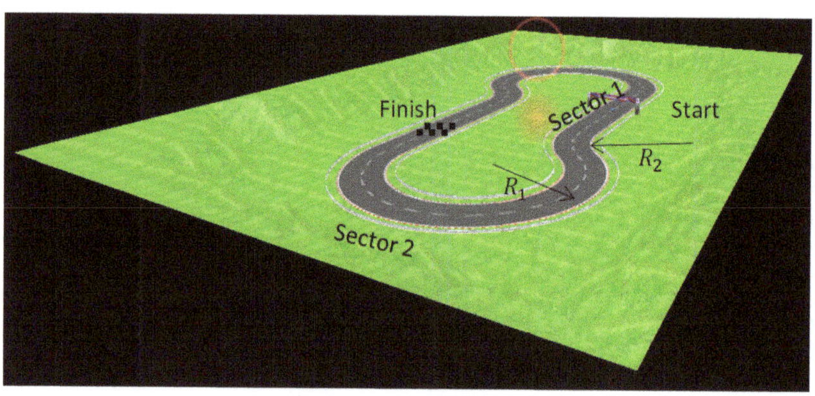

Figure 2. Hook-type test track.

Usually, during the first section of the route—straight—the driver accelerates and, at the maximum speed, begins the passage of the turn. The peculiarity of this rotation is that it does not contain a transition spiral curve part between the straight and the rotation sections. A spiral curve is a geometric element that can be added to a regular curve and provides a gradual transition (red part in Figure 3) from movement in a straight line (blue part in Figure 3) to movement along a circle (green part in Figure 3).

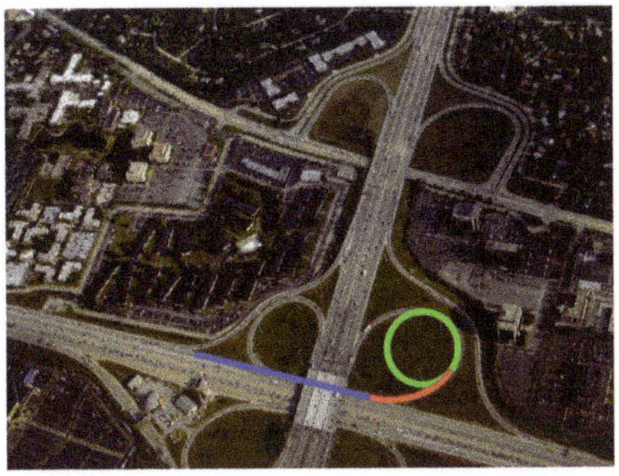

Figure 3. Rotation curve. The transition between the straight part (blue) and circular part (green) is a spiral curve (red). Picture is taken from https://cifrasyteclas.com/clotoide-la-curva-que-vela-por-tu-seguridad-en-carreteras-y-ferrocarriles/.

A sharp transition from a straight line to a turn results in stability losses, which, if the track is slippery, may cause an accident. The spiral provides a transition zone, where the driver slowly turns the steering wheel, lateral acceleration slowly increases when entering the spiral or slowly decreases when exiting the spiral and stability is not lost. Such spiral transitions were originally introduced on the railways for safety reasons. They were also implemented on highways in recent years. The mathematical form of the spiral varies [8]. One of the common forms is the Euler or clothoid spiral [9]. In India, the usual transition curve is a hyperbole third-order, and in Germany, autobahns are designed as a continuous series of linked clothoids without tangential sections or circular curves [10]. In the proposed study, the track did not have such transitions. We did this in order to achieve maximum generalization.

In our set of experiments, we did a car runs on the flat road without height changes. Parameters of the track are shown in Table 2. Also, for the experiments, we varied three types of slippery conditions and friction μ between the tires of the car and the surface of the track: rain (μ = 0.5), rain and snow (μ = 0.3), and ice (μ = 0.1).

Table 2. Main features of the test track.

Feature	Value
Total length, m	300
Lane width, m	20
Length of sector 1, m	90
Radius of turn 1, R_1 m	50
Length of sector 2, m	210
Radius of turn 2, R_2, m	50

2.3. Servo Control as a PD Controller

In our earlier studies, we showed that the canonical well-known servo control (Figure 4) is a variation of the PD controller. To avoid repetition, we give here only the main statements; the details can be found in our paper [6].

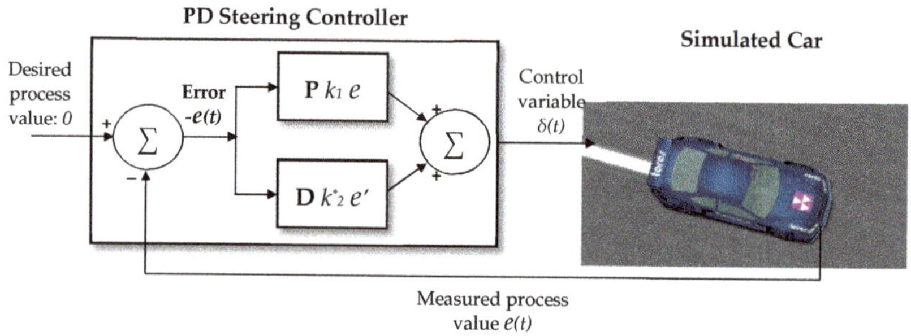

Figure 4. Servo-control model of steering as a PD steering controller. The steering angle function (SAF), defining the steering angle δ is implemented as a sum of the proportional-(P) and derivative (D) terms of the *error*—the deviation e from the center of the lane.

The steering angle defined by this model is a linear combination of scalable deviation from the desired trajectory parameters—the distance e and angle θ shown in Figure 5. The SAF (steering angle function) of the servo control model could be expressed as the following Equation (1):

$$\delta = k_1 e + k_2 \theta \tag{1}$$

where δ is the steering angle, and the optimal values of the scaling coefficients (gains) k_1 and k_2 have an impact to the main requirements to the steering [11]—smoothness, fast response, and stability in the way the car returns to the center of the lane after it deviates from it. These parameters could be chosen from the steering lock angle restrictions in 30^0 in different ways depending on the specific conditions and features of the road [4].

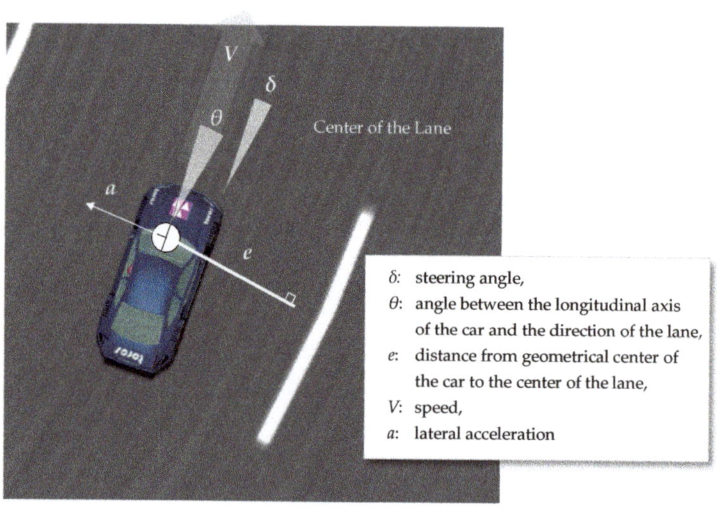

Figure 5. Most relevant variables pertinent to the state of the car. Variables θ and e are directly involved in the calculation of the steering angle δ according to the servo-control model.

For small angles θ and very short periods of time dt:

$$\theta \approx de/dx = de/(V\,dt) \tag{2}$$

For constant speed V, Equation (2) could be rewritten as:

$$\theta \approx k_v\, de/dt = k_v\, e' \quad (3)$$

Applying simple mathematical transformations, we can represent Equation (1) in the following shape expressed the servo-control model of steering as a PD controller:

$$\delta = k_1\, e + k_2\, (k_v\, e') = k_1\, e + k^*{}_2\, e' \quad (4)$$

where e' is the first derivative of the lateral deviation of the car from the center of the lane. From the PD controller point of view, servo control model represents by itself a closed-loop system. The input—measured *process value* is equal to the absolute value of the error *e*—the deviation of the car from the center of the lane. Its output—the *control variable*—the steering angle δ, is a sum of the proportional (P) and derivative (D) terms of the *error*. The controller attempts to minimize the value of the *error* by constantly adjusting the steering angle δ, which, presumably, would yield a trajectory as close as possible to the center of the lane.

2.4. Extending the Servo-Control Model: A PD Steering Controller with Prediction

The main common disadvantage of both PD controllers is lagging. It results in an even worse effect on the slippery road. In addition, many real-world physics effects are not taken into account in the controller equations. In practice, these disadvantages lead to the late entrance to the turn, attempts to return to desired trajectory, oversteering and disadvantageous position at the start of second turn, exiting, which could only result in consecutive course deviations, as demonstrated in Figure 6.

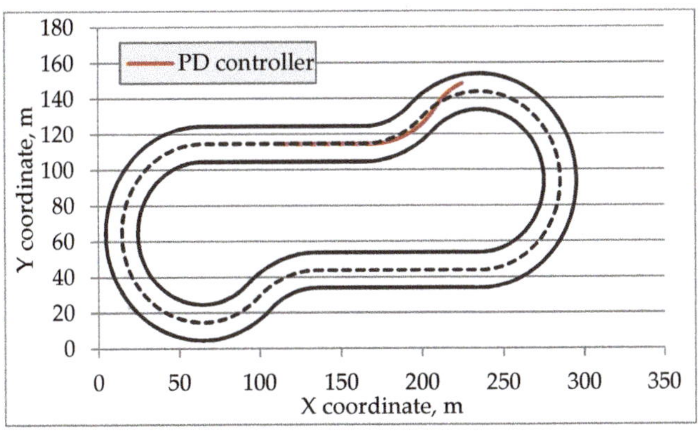

Figure 6. Tracking of the center of the car on the track during the race with friction 0.1 and 1.05 V_{CR}. In the highest point of the trajectory, a crash was occurring.

While the second disadvantage was already addressed in previous research [6], we did not manage to find sources regarding the first one.

The vehicle position could be predicted using current velocity (Figure 7; Figure 8):

$$\delta = k_1 e_{predicted} + k_2 \theta \quad (5)$$

where θ is the angle between the car direction and the road, and $e_{predicted}$ is the distance between the predicted car center and the road calculated by the following Equations (5)–(9):

$$e_{predicted} = F\left(x_{predicted}\,,\, y_{predicted}\right) \quad (6)$$

$$x_{predicted} = x_0 + V_x t = x_0 + Vt \cos\alpha \quad (7)$$

$$y_{predicted} = y_0 + V_y t = y_0 + Vt \sin\alpha \quad (8)$$

$$\alpha = \theta + A_{road} \quad (9)$$

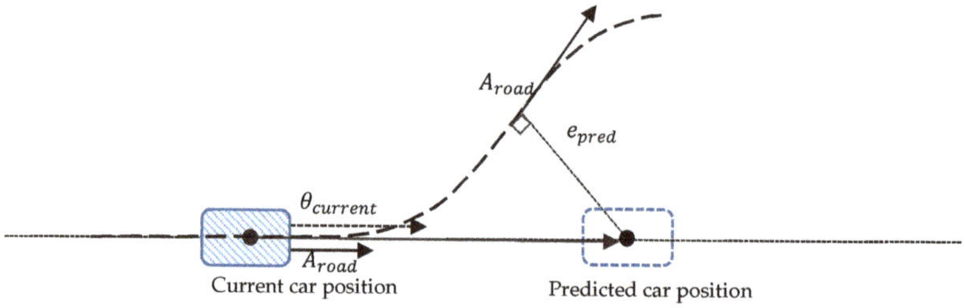

Figure 7. Predicting the lateral deviation of the car e_{pred}.

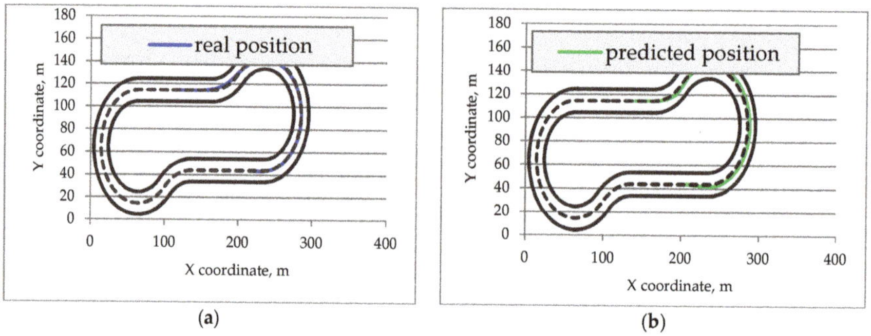

Figure 8. Real trajectory of the car (**a**) and predicted position of the car (**b**). Time of the prediction—1.5 s, $\mu = 0.1$.

Here, $x_{predicted}$ and $y_{predicted}$ are predicted coordinates of the position of the car, A_{road} is the angle of the road at the point that is closest to the car, F is the function calculating the distance between the car and the center of the lane, V, V_x, is the speed of the car, and its two orthogonal components, respectively, t. is the predicting time interval, and α is the angular deviation of the car from the center of the lane.

This controller we called Predictive PD controller (PPD controller).

It seems natural to also predict the yaw angle. However, it was not performed in our experiments. The reasons for this are not only atomicity of changes that facilitate development and analysis, but the aspiration to achieve human-like behavior of professional and experienced drivers. In contrast to the canonic controller, drivers often use the future position prediction for steering. According to the research [12], having to deal with a lot of information, the human brain experiences cognitive pressure with high speed, which results in a decrease of field of view. In such conditions, it has to operate only the prediction of the position, not the yaw angle. On the other hand, such limitations of human perception do not limit researches in the development of probably more precise models, which predicts both parameters. Nevertheless, such investigations are left for further research and, here, we were curious about how much a change of the value of one of the parameters will affect the quality of driving. We elaborate on this in the discussion section of the paper.

2.5. Steering Controller Obtained by Genetic Programming (GP)

To achieve a more thorough and complete analysis of the new controller, in this paper, we also presented the results of experiments conducted with the previously obtained model with an extended set of parameters and relaxed structure, evolved via a genetic programming (GP-RMEP) [6] method.

The GP approach allowed us to construct a controller of arbitrary complexity and structure, which showed results significantly superior to the canonical PD controller [6]. It took a lot of time at the stage of evolution to build a control equation that would demonstrate good results simultaneously in different environmental conditions, but the resulting model showed good results in terms of vehicle speed during the race and quality of the trajectory. However, as it turned out, this model also has its drawbacks; in particular, it produces frequent oscillations of the steering wheel, which, in the long term, can cause mechanical damage similar to what race cars get. Although this is not a critical flaw, it encourages us to continue research in this area. To avoid self-repetition, we present here in Table 3 only its main parameters, while our previous work contains the details [6].

Table 3. Parameters of genetic programming (GP).

Parameter	Value
Evolved individuals	SAF δ
Genetic representation	Parse tree
Set of non-terminals (functions)	{+, −, *, /}
Set of terminals	Variables pertinent to the state of the car, and their derivatives: lateral deviation (e, e'), speed (V), steering angle(δ), lateral acceleration (a, a') angular deviation (θ, θ'), and a random constant (C)
Population size	200 individuals
Selection	Binary tournament, ratio 0.1
Elitism	Best 4 individuals
Crossover	Single point, random, ratio 0.9
Mutation	Single point, random, ratio 0.05
Fitness value	Sum of (i) the area under the trajectory of the car around the center of the lane and (ii) the average of its lateral velocity.
Termination criteria	(#Generations > 200) or (no improvement of fitness during 16 consecutive generations)

2.6. Target Quality Function Evaluation

The target quality function value is intended to estimate the quality of the steering produced by the obtained SAF. We defined the criteria of such a quality from the desired characteristics of the trajectory of the car during the trial. It is simulated on a given test track (as shown in Figure 2) featuring a given friction coefficient μ, as follows: first, the simulated car is initially positioned at the starting position of the track. Then the car accelerates slowly to a given target speed. In order to render the task of controlling the car challenging, but solvable, the target speed was constant (maintained by simulated cruise control), and equal to 0.85, 0.9, and 0.95 of the critical speed V_{CR}. The critical speed V_{CR} is the speed at which the car theoretically could pass the turns of the test track with the given coefficient of friction without losing control, running off the track, and eventually crashing. This speed is approximated as the speed at which the centrifugal forces during a steady-state cornering [13] become theoretically equal to the friction force. At the traveling speed of 0.85 of V_{CR}, the car inherently suffers from intermittent instability (due to the yaw inertia both in the entry and exit of corners, and due to dynamic lateral weight transfer in corners [3,7]) that we intend to counter by the use of the obtained SAF. The car, traveling at V_{CR} (or above) is theoretically uncontrollable. Consequently, there would be no existing SAF that results in a steerable car. Similarly, a car traveling considerably slower than V_{CR} does not suffer from any instability, and its steering could be accomplished adequately by the canonical servo-control models. The speeds of the car during the trials on the track with different friction coefficients are shown in Table 4.

Table 4. Speed of the car during the trial on the test track with different friction coefficients.

#Road Condition	Friction of Tires, μt	Friction of Road Surface, μs	Overall Friction, $\mu = \mu t \times \mu s$	Critical Speed, V_{CR}, m/s	Speed of the Car (0.85 V_{CR}), m/s	Speed of the Car (0.9 V_{CR}), m/s	Speed of the Car (0.95 V_{CR}), m/s
1	1.0	0.5 (rainy)	0.5	15.65	13.3	14.2	15
2	1.0	0.3 (icy and snowy)	0.3	14	10.4	11	11.6
	1.0	0.1 (icy)	0.1	12.12	6	6.3	6.7

The speed of the car is kept constant during the trial by a simple, handcrafted cruise control mechanism that maps the difference between the desired speed (e.g., as shown in Table 4, 13.3 m/s for the trial on a track with friction coefficient μ = 0.5) and the actual one into an increment (or decrement) of the position of accelerator pedal. As the car reaches the desired speed, the steering of the car is assumed by the obtained SAF. Then, the latter starts to continuously (with a sampling frequency of 40 Hz) produce the desired steering angle δ calculated for the currently perceived values of the parameters pertinent to the state of the car. The desired trajectory of the car is the center of the lane. The estimation of the current trajectory obtained via the SAF with new parameters is calculated with the same target quality function as earlier [4] with the purpose of a correct result comparison.

According to this, the target quality function F is a weighted sum of the following two components: (i) the area A_T under the trajectory of the car around the center of the lane (as an integral of the absolute value of lateral deviation e) and (ii) the average of the lateral velocity V_{L_AVR} (as an integral of the lateral acceleration a) of the car:

$$F = A_T + C_V\, V_{L\_AVR} \tag{10}$$

We would like to note that for different steering tasks, we might need to keep track of both components of the target quality function of obtained SAF separately (and to implement a two-objective optimization [14]) instead of fusing both these components in a single scalar value. This would allow us to obtain a set of Pareto-optimal SAF that features different combinations of the area under the trajectory of the car and the average of its lateral velocity. SAF featuring a wide area under the trajectory might be needed in a slow and comfortable lane change on a low-traffic highway. On the other hand, the SAF that results in oscillating trajectories with higher lateral accelerations might be needed to safely circumnavigate suddenly appeared obstacles. However, for the given task, the proposed evaluation of the target quality function is sufficient.

3. Experimental Results

For each road condition described in Table 4, we developed a new analytic equation and tested them with different levels of target speed. We pick optimal parameters for a PD controller according to algorithms in our previous work [6] to optimize its quality function and controller perception. So, we ran simulations for predictive PD, best PD, and, constructed via genetic programming [7], GP-RMEP controlling models. Algorithms were designed for the same track with the same conditions and the same estimation quality function F, so we could compare them. Resulting car trajectories are presented in Figure 9.

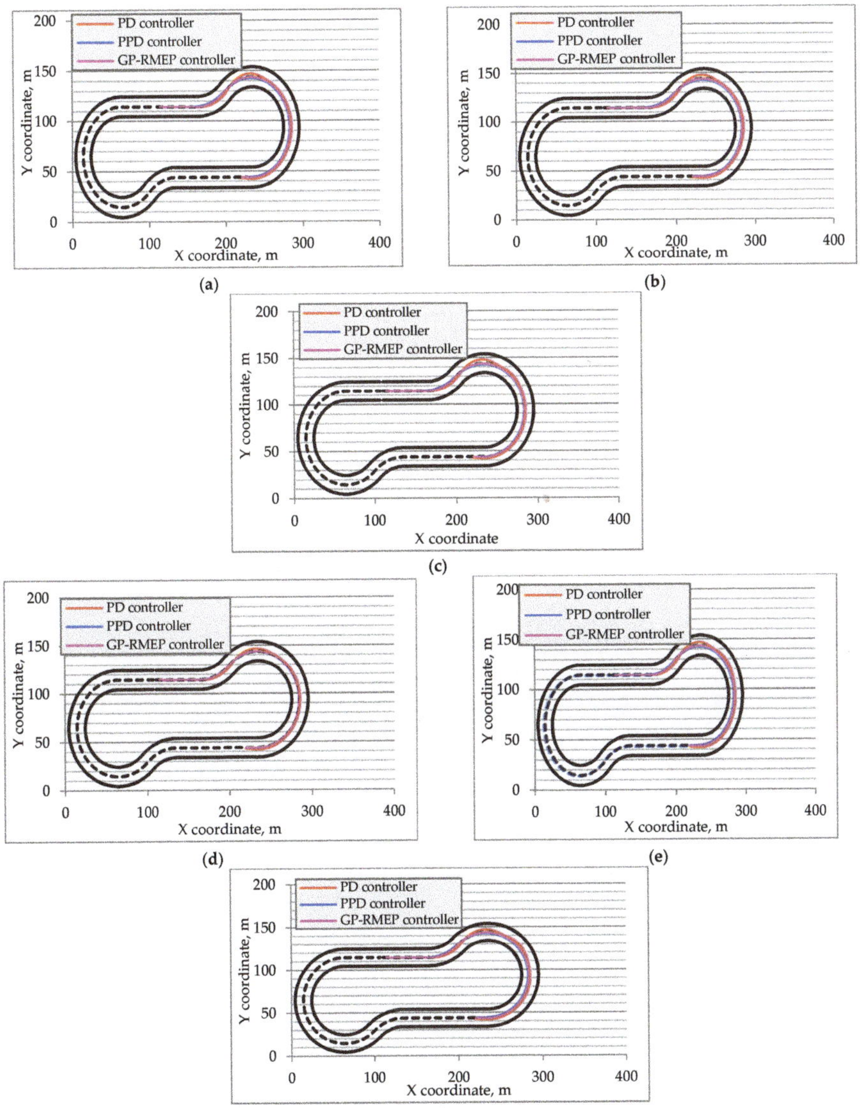

(**a**) (**b**) (**c**) (**d**) (**e**) (**f**)

Figure 9. *Cont.*

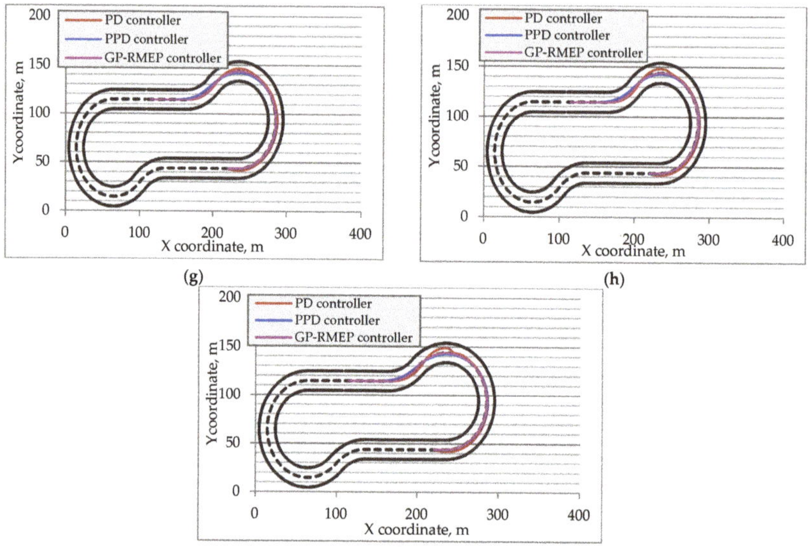

(g) (h) (i)

Figure 9. Car trajectories on the track tuned with prediction SAF of standard PD controller for friction coefficients. The purple curves correspond to the GP-RMEP controller, blue curves to the trajectories controlled by SAF with prediction, red curves to the original SAF. (**a**) μ 0.5, 0.85 V_{cr}; (**b**) μ 0.5, 0.9 V_{cr}; (**c**) μ 0.5, 0.95 V_{cr}; (**d**) μ 0.3, 0.85 V_{cr}; (**e**) μ 0.3, 0.9 V_{cr}; (**f**) μ 0.3, 0.95 V_{cr}; (**g**) μ 0.1, 0.85 V_{cr}; (**h**) μ 0.1, 0.9 V_{cr}; (**i**) μ =0.1, 0.95 V_{cr}.

As Figure 9 illustrates, under each road condition, the trajectory provided by the PD controller version with prediction was better (i.e., closer to the center of the lane). In Figure 10, the changes of steering angle for parameters $\mu = 0.3, 0.95V_{cr}$ is shown. It could be noted that the new controller provides smooth steering, similar to the PD controller but with lower amplitude. This characteristic also corresponds to driving stability. We analyzed the smoothness of each function as a number of sign changes in the approximate first derivative: $d = \frac{\Delta\alpha}{\Delta t}$. The results were as follows: GP-RMEP—349 changes, PD controller—218 changes, PPD controller—98 changes. This PPD controller is more stable than PD and has lower angle change amplitudes. As for comparison with the GP method, the high frequency of oscillations makes them less influencing but exhausts the tires and mechanical parts of the steering. We compared their target quality function values, and the results are shown in Table 5.

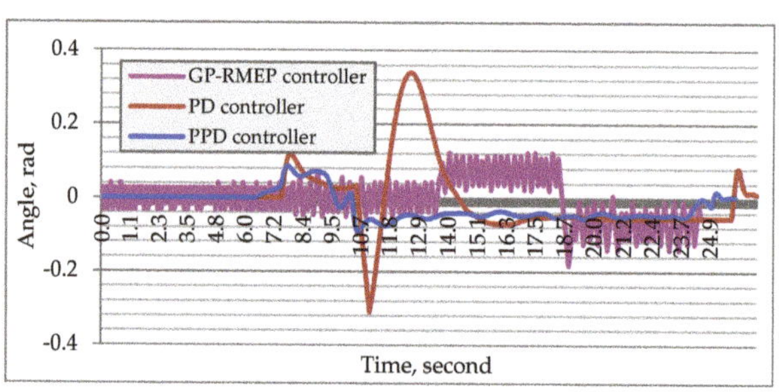

Figure 10. Dynamics of the steering angle for different types of controllers—the controller input is $\mu = 0.3, 0.95V_{cr}$.

Table 5. Steering controllers target quality function for each parameter combination.

#Road Condition	Overall Friction μ	0.85 of Critical			0.9 of Critical			0.95 of Critical		
		PD	PPD	GP-RMEP	PD	PPD	GP-RMEP	PD	PPD	GP-RMEP
1	0.5	685	298	373	711	334	318	843	364	385
2	0.3	1693	383	374	1801	417	399	1854	458	413
3	0.1	1659	408	381	1717	432	420	1782	471	461

As the results shown in Table 5 demonstrate, the PPD controller outperforms the PD and is comparable with the GP-RMEP controller in terms of target quality function. Thus, the new controller has a trajectory close to the best trajectory obtained with the GP controller but does not have oscillations that could lead to mechanical damage [15]. Since, according to our earlier studies [6,13], oscillations in the GP-RMEP method are part of its tactics aimed at increasing and maintaining the largest slip angle on tires and, accordingly, cannot be removed from the method without reducing the speed and safety of the car, the result we obtained, which is close in quality to GP-RMEP and free from oscillations, is an important result for us.

Also during the analysis of the proposed model, we discovered several features, as described below.

3.1. Time Needed to Return on Desired Trajectory

The first feature is a more rapid return to the center lane. As shown in Table 6, in each considered case, the time needed to return on the desired trajectory decreased by 4% to 11%, which corresponds to 10-20 m of movement with speeds from Table 4 and could be crucial for safety.

Table 6. Times needed to return to the lane center, second.

#Road Condition	Overall Friction μ	0.85 of Critical			0.9 of Critical			0.95 of Critical		
		PD	PPD	GP-RMEP	PD	PPD	GP-RMEP	PD	PPD	GP-RMEP
1	0.5	14.72	13.71	14.18	14.56	13.57	12.81	14.44	13.35	13.52
2	0.3	18.47	17.59	17.18	18.3	17.89	17.19	18.19	17.1	16.9
3	0.1	37.51	34.58	33.14	35.59	32.17	31.42	34.31	30.48	30.16

These times demonstrate how fast the vehicle returns from the position occupied while being affected by the forces during the turn. The parameter is directly related to the safety of a driver. Large values could correspond to both a noticeable distance to desired trajectory or inconvenient vehicle orientation. Both cases may cause the following complications, resulting, for instance, in a turning vehicle in oncoming traffic.

3.2. Critical Speed Rising

Another feature of the proposed approach, which improves safety, is demonstrated in Figure 11. Here, with identical environment parameters, the vehicle without prediction, moving at a speed above critical, loses control and crashes, while the model with prediction successfully finishes the race. The reason for this results from differences in trajectories. As it was shown in previous studies [6], the vehicles' critical speed during the turn is proportional to $\sqrt{R}$. The model with prediction starts the turn earlier, which results in increased R, and as a result, increased critical speed. In target quality function, this should lead to improvements in the parameter, corresponding to the car stability—second addend in Equation (10). This is covered in the Discussions Section in Table 7.

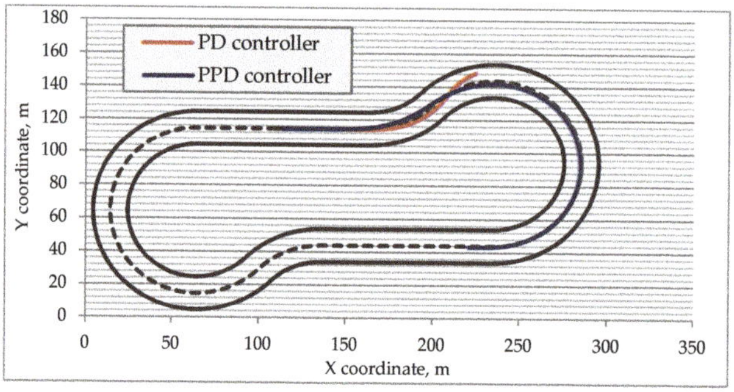

Figure 11. Car trajectory on the track tuned with prediction SAF of standard PD controller for friction coefficient $\mu = 0.1$ with target speed equal to $1.05V_{CR}$. The blue curve corresponds to the trajectory controlled by PPD controller, and red to the original PD controller.

Table 7. Distance to the obstacle for each parameter combination, meter.

#Road Condition	Overall Friction μ	0.85 of Critical			0.9 of Critical			0.95 of Critical		
		PD	PPD	GP-RMEP	PD	PPD	GP-RMEP	PD	PPD	GP-RMEP
1	0.5	8.88	12.14	12.33	8.73	11.81	12.07	8.34	11.4	11.84
2	0.3	8.09	10.81	11.6	7.93	10.79	11.38	7.91	10.74	11.14
3	0.1	8.03	10.87	11.23	7.89	10.8	11.16	7.86	10.07	10.9

3.3. *Safe Distance*

The third feature is increased distance to potential obstacle during the turn, i.e., the moment of the minimal stability of a car (Figure 12).

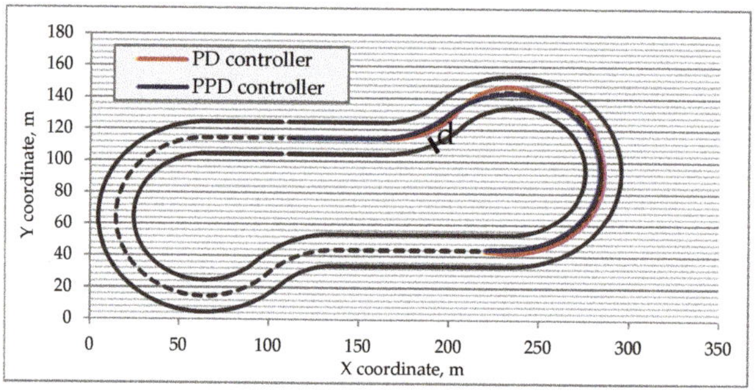

Figure 12. Distance between the car and obstacle (road turn) that caused the turn friction coefficient $\mu = 0.3$ with target speed equal to $0.95V_{CR}$.

Due to prediction, the vehicle starts the turn earlier. This not only makes the trajectory smoother and increases turn radius but also allows avoiding potential causes of turn—road twist or unexpected obstacle. According to data from Table 7, the average distance to an obstacle increased by 35%.

These numbers, as well as the values of the target quality function, indicate the remoteness of the results occurred by each method from the desired ones (if they were equal to zero, the car would move along the chosen trajectory without deviations caused by instability). In other words, these numbers could be interpreted as the error amount of methods. Figure 13 demonstrates distance and angle errors for all compared methods. In these terms, PPD performs better than PD and has similar changes of amplitudes with the GP-RMEP controller.

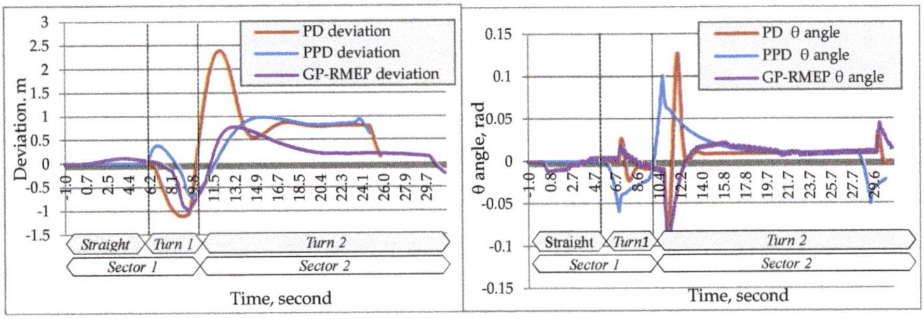

Figure 13. Distance error (left picture) and yaw angle error (right picture) with 0.95 V_{cr} and 0.3μ.

4. Discussion

Data from Table 8 demonstrate that improvements of a new model affected both target quality function addends—deviation from the trajectory and lateral acceleration. An increase of one of the parameters may lead to the decrease of another. The closer the vehicle is to the center of the lane, the stronger the forces affecting it during the turn. However, in this case, the new model demonstrated improvements in both parameters, so called non-zero sum, which indicates qualitative model enhancement in contrast to parameter tuning.

Table 8. Steering controllers target quality function for each parameter combination split by addends corresponding to deviation from the center of the lane and lateral acceleration, respectively.

#Road Condition	Overall Friction μ	0.85 of Critical Speed		0.9 of Critical Speed		0.95 of Critical Speed	
		PD	PPD	PD	PPD	PD	PPD
1	0.5	239 + 446	79 + 219	255 + 456	103 + 231	288 + 555	110 + 254
2	0.3	779 + 914	186 + 196	823 + 978	215 + 201	870 + 984	252 + 206
3	0.1	1241 + 418	346 + 62	1291 + 426	367 + 65	1305 + 477	405 + 66

Another issue that was not covered previously is the value of prediction time distance. In other words, it is the task of finding the optimal parameter for Equations (7)–(9). Previously, the value was chosen manually for each configuration and was in the range [0.8 ... 1.8] seconds. Too short of a prediction time does not allow achieving the effect described in Section 2.4 and leads to trajectory Type (1) in Figure 14. In contrast, too long of a period provides changes to the trajectory, which makes it too far from that desired, as shown on Trajectory (3), Figure 14.

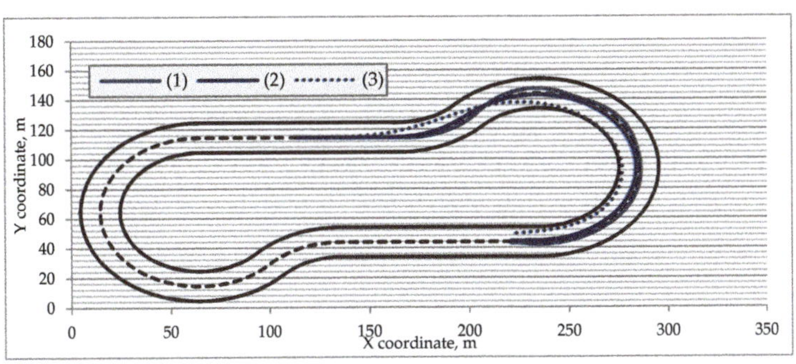

Figure 14. Different types of the car trajectory behavior depending on the prediction time. From (1) to (3), this duration becomes longer.

Therefore, we ran a series of experiments in order to find the optimal value of this parameter. Figure 15 shows the results. We started the search from the small prediction times provided trajectory (1) and increased the time value until the result become equal to 1000 (which means the car crashed). The optimal values of prediction time could not be less than what we start from because they will turn into the PD controller results, and they could not be more than those that lead to the car crash already because of its remoteness from reality. We also note that an increase of target speed along with a decrease of friction coefficient increases the optimal prediction time. In other words, under less stable conditions, the prediction should be farther.

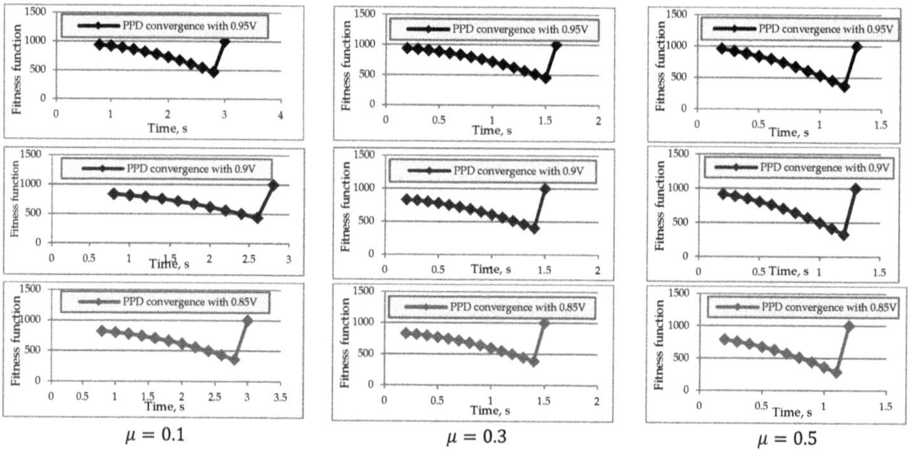

Figure 15. Target quality function convergence of the PPD controllers under the different speed levels and friction values (μ = 0.1, 0.3, 0.5) depends on time prediction.

Another thought that we mentioned before in the Introduction Section is a special form of the safe trajectory that has a linear curvature profile, which is comfortable for the turning car with high speed.

In order to compare the obtained by PPD controller trajectory shape with clothoid, we combined them on a single plot. Clothoid could be constructed by solving the system of differential equations below, called the "reconstruction equation".

A clothoid is uniquely defined by:

- Coordinates and heading at which it starts: (x_0, y_0, θ_0);
- Its length L;
- Its linear curvature function, which is determined by the two coefficients (k_0, k_1).

With these five values $(x_0, y_0, \theta_0, L, k_0, k_1)$, we can evaluate the clothoid's position and heading $(x(s), y(s), \theta(s))$ at any point s in area $[0; L]$. We did that by solving the following equations:

$$x'(s) = cos\theta(s)$$

Now, in Figure 16, we can see that our new model produces a trajectory that has a very similar shape to the clothoid (0, 0, 0, 100, 0, 0.1), which was turned around the center.

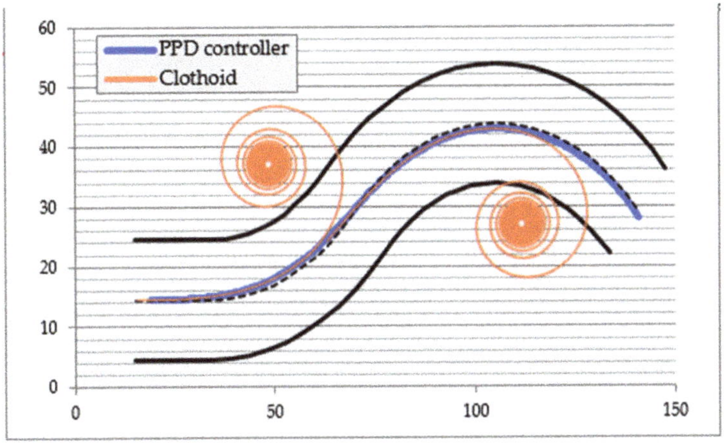

Figure 16. Trajectory of the first left and right turns combined with clothoid spiral. Both turns demonstrate a gradually increasing turning radius.

According to the researches [16], the best and smoothest transition curve to be used as a section of the path is the clothoid. The fact that the PPD controller we developed approached the same conclusion with its results, along with their comparison with the results of other controllers, once again indirectly indicates the effectiveness of the new controller.

5. Conclusions

We researched and implemented the predictive PD controller in the TORCS simulator environment for the race car model (Mercedes CLK). In spite of the fact that we tested a car with specific parameters (a race car), providing, among other things, greater stability (for example, the height of its center of gravity is lower than usual), some of the found features and control tactics can be applied to regular cars. The height of the center of the gravity—together with the lateral forces—would determine the amount of the lateral weight transfer of the car on cornering, which, in turn, would affect the distribution of the normal forces on the tires. On slippery roads, due to the lower lateral forces applied to the cornering car (due to the lower friction coefficients), we assume that the lateral weight transfer would be negligible, regardless of the height of the center of the gravity of the car. This controller relies on the predicted position of a vehicle instead of the current position. The several series of experiments confirmed the relevance of the applicability of such an approach in practice. In fact, the new model demonstrated not only better results but also more native and adequate behavior, which provides greater safety and stability of driving on a slippery road. Based on the reaction style adopted by the human driver to the obstacle that appears in the field of view, this model is able to avoid them by changing its behavior in advance. Depending on the selected speed, the optimal prediction time was computed. During the analysis of the method, it was compared with previously studied methods, which we referred to in this article (PD, PID, GP-RMEP). During the comparison, it turned out that our modification of the PD controller showed results close to those found using genetic programming (GP-RMEP), which demonstrated best adaptability and applicability on slippery roads. The additional studies of this parameter, involving machine learning, are left for further research.

Author Contributions: Conceptualization, I.T.; Investigation, N.A.; Methodology, N.A.; Project administration, K.S.; Software, N.A.; Supervision, I.T. and K.S.; Validation, I.T.; Writing—original draft, N.A.; Writing—review & editing, I.T. and K.S. All authors have read and agreed to the published version of the manuscript.

Funding: This work was funded in part by MEXT-supported Program for Strategic Research Foundation at Private Universities in Japan (2014–2018).

Conflicts of Interest: The authors declare no conflict of interest.

References

1. Stone, P.; Brooks, R.; Brynjolfsson, E.; Calo, R.; Etzioni, O.; Hager, G.; Hirschberg, J.; Kalyanakrishnan, S.; Kamar, E.; Kraus, S.; et al. Artificial Intelligence and Life in 2030. In Proceedings of the One Hundred Year Study on Artificial Intelligence: Report of the 2015-2016 Study Panel, Stanford, CA, USA, 6 September 2016; p. 52.
2. Le-Anh, T.; De Koster, M.B. A Review of Design and Control of Automated Guided Vehicle Systems. *Eur. J. Oper. Res.* **2006**, *171*, 1–23. [CrossRef]
3. Cheein, F.; De La Cruz, C.; Bastos, T.; Carelli, R. SLAM-based Cross-a-Door Solution Approach for a Robotic Wheelchair. *Int. J. Adv. Robot. Syst.* **2010**, *7*, 155–164. [CrossRef]
4. Alekseeva, K.S.N.; Tanev, I. Evolving a Single-variable Controller for Automated Steering of a Car on Slippery Roads. In Proceedings of the SICE Annual Conference 2018: SCIE 2018, Nara, Japan, 11–14 September 2018; pp. 680–685.
5. Huang, J.; Tanev, I.; Shimohara, K. Evolving a general electronic stability program for car simulated in TORCS. In Proceedings of the 2015 IEEE Conference Computational Intelligence and Games (CIG), Tainan, Taiwan, 31 August–2 September 2015; pp. 446–453.
6. Alekseeva, N.; Tanev, I.; Shimohara, K. Evolving the Controller of Automated Steering of a Car in Slippery Road Conditions. *Algorithms* **2018**, *11*, 108. [CrossRef]
7. Wymann, B.C.R.; Dimitrakakis, C.; Sumner, A.; Espié, E.; Guionneau, C. TORCS, the Open Racing Car Simulator, v1.3.5. 2013. Available online: http://www.torcs.org (accessed on 20 February 2020).
8. Komoriya, K.; Tanie, K. Trajectory Design and Control of a Wheel-type Mobile Robot Using B-spline Curve. In Proceedings of the IEEE/RSJ International Workshop on Intelligent Robots and Systems '(IROS '89)' the Autonomous Mobile Robots and Its Applications, Tsukuba, Japan, 4–6 September 1989; pp. 398–405.
9. Marzbani, H.; Jazar, R.; Fard, M. Better Road Design Using Clothoids. In *Sustainable Automotive Technologies*; Springer: Cham, Switzerland, 2015; pp. 25–40.
10. Lamm, R.; Psarianos, B.; Choueiri, E.M. A practical safety approach to highway geometric design International case studies: Germany, Greece, Lebanon, and the United States. In Proceedings of the International Symposium on Highway Geometric Design Practices, Boston, MA, USA, 30 August–1 September 1995.
11. Melder, N.; Tomlinson, S. Racing Vehicle Control Systems using PID Controllers. *Game AI Pro* **2014**, 491–500.
12. Zeeb, K.; Buchner, A.; Schrauf, M. Is take-over time all that matters? The impact of visual-cognitive load on driver take-over quality after conditionally automated driving. *Accid. Anal. Prev.* **2016**, *92*, 230–239. [CrossRef] [PubMed]
13. Alekseeva, N.; Tanev, I.; Shimohara, K. On the Emergence of Oscillations in the Evolved Autosteering of a Car on Slippery Roads. In Proceedings of the 2019 IEEE/ASME International Conference on Advanced Intelligent Mechatronics (AIM), Hong Kong, China, 8–12 July 2019; pp. 1371–1378.
14. Coello, C.A.C.; Lamont, G.B.; Van Veldhuizen, D.A. *Evolutionary Algorithms for Solving Multi-Objective Problems*; Springer: New York, NY, USA, 2007.
15. Frère, P. *Sports Car and Competition Driving*; Pickle Partners Publishing: Auckland, New Zealand, 2016.
16. Marzbani, H.; Simic, M.; Fard, M.; Jazar, R. Better Road Design for Autonomous Vehicles Using Clothoids. In *Intelligent Interactive Multimedia Systems and Services*; Springer: Cham, Switzerland, 2015; pp. 265–278.

algorithms MDPI

Article

Neural PD Controller for an Unmanned Aerial Vehicle Trained with Extended Kalman Filter

Javier Gomez-Avila , Carlos Villaseñor, Jesus Hernandez-Barragan, Nancy Arana-Daniel *, Alma Y. Alanis and Carlos Lopez-Franco

Centro Universitario de Ciencias Exactas e Ingenierías, Universidad de Guadalajara, Blvd Marcelino García Barragán 1421, Guadalajara 44430, Mexico; jenrique.gomez@academicos.udg.mx (J.G.-A.); cavp@outlook.com (C.V.); jesus.hdez.barragan@gmail.com (J.H.-B.); almayalanis@gmail.com (A.Y.A.); carlos.lopez@cucei.udg.mx (C.L.-F.)

* Correspondence: nancyaranad@gmail.com; Tel.: +52-331-547-3877

Received: 19 December 2019; Accepted: 12 February 2020; Published: 18 February 2020

Abstract: Flying robots have gained great interest because of their numerous applications. For this reason, the control of Unmanned Aerial Vehicles (UAVs) is one of the most important challenges in mobile robotics. These kinds of robots are commonly controlled with Proportional-Integral-Derivative (PID) controllers; however, traditional linear controllers have limitations when controlling highly nonlinear and uncertain systems such as UAVs. In this paper, a control scheme for the pose of a quadrotor is presented. The scheme presented has the behavior of a PD controller and it is based on a Multilayer Perceptron trained with an Extended Kalman Filter. The Neural Network is trained online in order to ensure adaptation to changes in the presence of dynamics and uncertainties. The control scheme is tested in real time experiments in order to show its effectiveness.

Keywords: PD controller; multilayer perceptron; extended kalman filter

1. Introduction

Unmanned Aerial Vehicles (UAVs) have become very popular thanks to recent progress in propulsion technologies and small sensors with low power consumption [1]. These kinds of vehicles surpass other types of robotics platforms in both military [2] and civilian applications [3], including surveillance, agriculture, traffic monitoring, fire detection and high social impact activities, such as search and rescue in disaster zones. Figure 1 shows the classification of several types of aerial vehicles based on [4]. This paper is focused on Vertical Take-Off and Landing (VTOL) vehicles, such as multirotors.

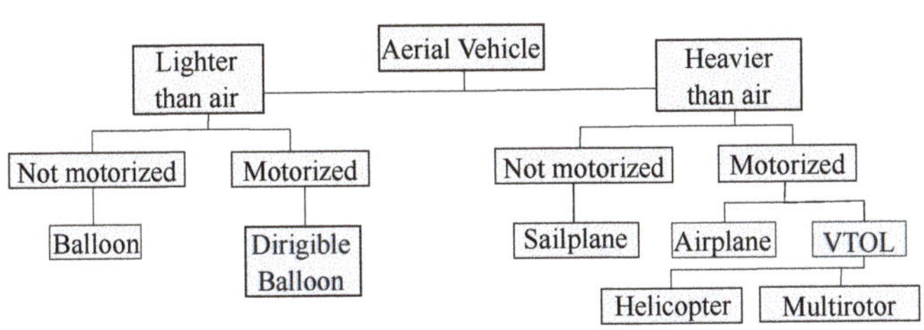

Figure 1. Aerial vehicles classification. This work is focused on multirotors.

VTOL vehicles are more cost-efficient than bigger drones like Medium Altitude Long Endurance (MALE) or High Altitude Long Endurance (HALE) and present the ability to hover above a reference position [5]. In contrast, multirotors have limited energy consumption and payload; consequently, light and compact sensors are required for the navigation, and in most of the cases, inertial sensors are not enough to obtain some states of the system, such as its position. Usually, multirotors are teleoperated by a ground station with a limited operational range, but when autonomous tasks are required, the positional feedback is crucial to control the UAV [5].

Drones are equipped with a Global Positioning System (GPS) to solve the problem of the estimation of the position. However, depending on the accuracy of the assignment, a GPS sensor may not be suitable; besides, the GPS signal is lost when working in indoor environments. For indoor flight control, the positional feedback is commonly carried out by motion capture systems, which each consist of a set of fixed cameras in a room. Unfortunately, approaches like this require previous knowledge of the scene and assembly and calibration of the motion capture system, which, in practice, would not be possible in search and rescue tasks. Generally, a combination of visual and inertial information is used to solve the problem of Simultaneous Localization and Mapping (SLAM) [6]. In this paper, a vision sensor is used; cameras are compact and lightweight sensors with low power consumption. These characteristics make them suitable for drones flying in unknown, GPS-denied environments.

Once the SLAM problem has been solved, it is possible to control the position of the vehicle. Proportional-Integral-Derivative (PID) controllers are widely applied in industry because of their simplicity, and they are usually used to control these kinds of UAVs. Nevertheless, multirotors are highly nonlinear, underactuated systems with six Degrees of Freedom (Dof) and four control inputs—torque in x, y and z, and thrust—and therefore, they are difficult to control with conventional methods [7,8]. To overcome the limitations of conventional controllers, direct control using a neural network is proposed. In this work, a Multilayer Perceptron (MLP) is implemented to adapt the gains of a PD controller. As reported in [9], Artificial Neural Networks (ANN) have shown satisfactory results when controlling nonlinear systems, and despite the limitations of conventional controllers, the approach presented in this paper can cover most of the disadvantages of PID, even if the multirotor changes its parameters during the flight.

The MLP is trained with the Extended Kalman Filter (EKF). The Kalman filter is an optimal estimator which deduces parameters based on noisy measurements. Its solution is recursive; therefore, the filter processes data as soon as it arrives and predicts the next value without the need for having the complete data set of observations [10]. This feature makes it faster and convenient for online applications, in contrast with other methods such as batch processing [11]. The idea behind the use of the EKF to train the ANN is that other training algorithms, such as gradient descent, recursive least squares and backpropagation, are particular cases of the Kalman filter; for this reason, the EKF is suitable for training [12,13]. When using the EKF for training, the weights of the neural network are the states that the filter estimates, and the output of the neural network is the measurement used by the Kalman filter. Then, the training of the ANN can be seen as an optimal filtering problem. The EKF has been successfully applied to estimate parameters of an ANN in [14–16].

Although several PID tuning algorithms have been proposed, those approaches are mostly offline applications [17,18] or simulation-only experiments [19]. The main contribution of this work is the control of both position and orientation of a quadrotor in real-time experimental tests using direct control; i.e., the output of the neural network is the control action for the UAV. The neural networks are trained online in order to adapt to changes in the dynamics and uncertainties. The objective is to find a robust solution to real applications [20] in which UAVs are capable of grabbing or deliver objects. This approach also uses a solution for the SLAM problem, to control the position using only onboard sensors, making it able to fly in unknown environments.

The remainder of this paper is organized as follows: in Section 2, some previous and related works are presented. The dynamic model of the platform used is described in Section 3. Section 4 presents the architecture of the neural network trained with EKF. In Section 5, the algorithm of localization is described. The quadrotor control scheme is shown in Section 6. Finally, simulation and experimental results are shown in Section 7. The conclusions of this work are discussed in Section 8.

2. Related work

Multirotors are commonly controlled by PID [21]. PID controllers have been widely used in industry, mainly because of the trade-off between efficiency and simplicity [22], and numerous offline tuning techniques have been reported in the literature [19]. The main problem with conventional approaches is that physical systems present parametric changes and uncertainties, and they are perturbed by external disturbances; consequently, an online, continuous tuning approach is needed.

Another common approach is the Linear Quadratic Regulator (LQR) for attitude stabilization [23–25]. While this may be applicable for some configurations, multirotors are non-linear systems that present uncertainties, such as unknown physical parameters, actuator degradation, unknown delays in-process communication and unmodeled dynamics [26]. Hence, an approach considering these characteristics must be used [27,28]. Nonlinear control techniques provide better performance [29], and one of the most applied methods is feedback linearization, which relies on cancellation of all nonlinearities to convert the non-linear system into a system with linear dynamics [30–32]. In [33–35], the authors use a backstepping control, which is designed to stabilize the whole system based on the Lyapunov stability theory, showing good results. However, backstepping requires the analytic calculation of the partial derivatives of the stabilizing functions, which becomes impractical as the order of the system grows [36], and in general, most of the nonlinear techniques require complete knowledge of the nonlinearities present in the plant and are vulnerable to modeling errors or parametric uncertainty [29]. In this paper, an adaptive controller based on an Artificial Neural Network (ANN) is proposed. The ANN has been used to control complex nonlinear systems [9,37–39]. The controller used in this scheme has the same simple structure and easy implementation of a PD controller but with the adaptability and learning capabilities of a neural network [9]. The system will be able to adapt to actuator faults, such as loss of effectiveness [40] and solve the principal disadvantages of traditional PID [41].

On the other hand, there is no onboard sensor to read the absolute position of the UAV. There are two common solutions to solve this; the first one consists of an external motion detection system, which has to be mounted, thereby limiting the applications to known indoor environments.

The second approach is based on solving the SLAM problem. For this, different sensors can be used, such as laser range scanners [42], stereo vision systems [43,44], RGB-Depth sensors [45–47] and monocular cameras [48–50]. In this work, a monocular vision system is preferred because of its lower power consumption and compact size, compared to the amount of information it delivers. The advantage is that the range of a camera is virtually unlimited [6], making it possible flying in both small and large environments. Despite the advantages of monocular vision, it is impossible to determine the scale of the environment using only one view, and it is necessary to fuse this information with inertial information provided by an Inertial Measurement Unit (IMU). To solve this, Parallel Tracking and Mapping (PTAM) for robot localization (introduced in [6,51,52]) will be implemented.

3. Multirotor Dynamic Model

The multirotor used for this work was the quadcopter. There are two principal configurations for quadcopters; in this case, the configuration selected is described in Figure 2.

The robotic platform where the algorithm was tested is the Parrot AR.Drone 2.0® quadrotor. Multirotor systems have an even number of rotors divided into two groups rotating in opposite directions. The configuration of the selected platform is depicted in Figure 2. For this specific configuration, where the robot structure does not match with the x and y axis from the body frame (Figure 3), the movement of the vehicle is given by the following combination of rotor actions: increasing or decreasing the speed of the four rotors in the same proportion changes the altitude of the system. Then, because the multirotor is an underactuated system, there is no actuator that generates movement in x and y directions directly; instead, these displacements are achieved by changes in the attitude due to combinations of the two pairs of propellers: the rotation in y axis (pitch θ) results in translational movement in x; contrarily, a rotation in x (roll ϕ) results in translational movement in y. Similarly, the orientation (yaw ψ) needs a combination of the four propellers and it is the result of the difference of the counter-torque between both pairs.

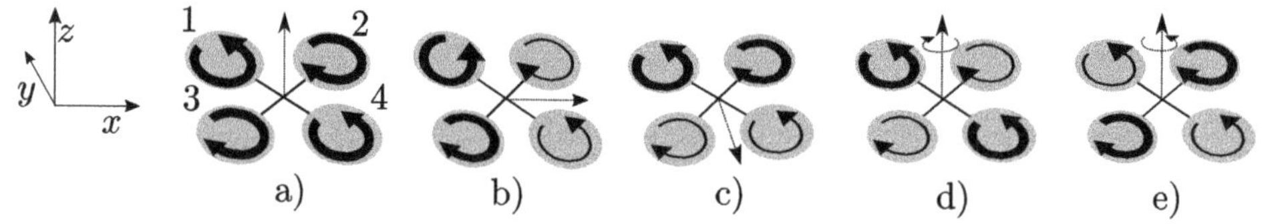

Figure 2. Illustration of the concept of the motion of a quadrotor. Let the propeller rotation speed be proportional to the width of the arrow on the propeller; the movement of the UAV is denoted with the dashed arrow at the center of the quadrotor. **(a)**Upward movement, **(b)** horizontal movement, **(c)** downward movement, **(d)** left turning movement and **(e)** right turning movement.

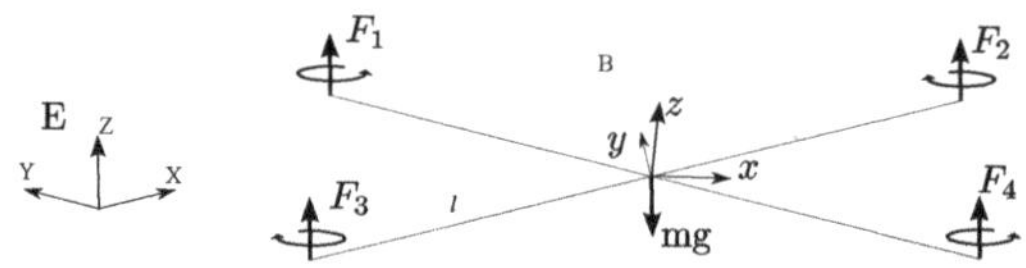

Figure 3. Quadrotor configuration. B represents the quadrotor fixed frame and E the inertial frame.

In general, the center of mass is considered to be at the origin of the body fixed frame. The quadrotor orientation in space is given by a rotation matrix $\mathbf{R} \in SO(3)$ from the quadrotor **B** to the inertial frame **E**. As in [4], the dynamics of the quadrotor may be expressed as follows

$$\begin{bmatrix} m\mathbf{I}_{3\times 3} & 0 \\ 0 & \mathbf{I} \end{bmatrix} \begin{bmatrix} \dot{\mathbf{V}} \\ \dot{\omega} \end{bmatrix} + \begin{bmatrix} \omega \times m\mathbf{V} \\ \omega \times \mathbf{I}\omega \end{bmatrix} = \begin{bmatrix} \mathbf{F} \\ \tau \end{bmatrix} \tag{1}$$

where $\mathbf{I}$ is the inertia matrix, $\mathbf{V}$ is the body linear speed vector and ω is the angular speed.

The quadrotor equations of motion can be expressed as

$$\begin{aligned} \dot{\zeta} &= \mathbf{v} \\ \dot{\mathbf{v}} &= -ge_3 + \mathbf{R}_{e_3}\left(\tfrac{b}{m}\sum \Omega_i^2\right) \\ \dot{\mathbf{R}} &= \mathbf{R}\hat{\omega} \\ \mathbf{I}\dot{\omega} &= -\omega \times \mathbf{I}\omega - \sum \mathbf{J}_r\,(\omega \times e_3)\,\Omega_i + \boldsymbol{\tau}_a \end{aligned} \tag{2}$$

where the gravity g is acting on z axis (e_3), $\dot{\zeta}$ is the linear velocity vector, the rotation matrix is represented by $\mathbf{R}$, $\hat{\omega}$ is the skew symmetric matrix of the vector of angular velocity and Ω_i represents the speed of rotor i. $\mathbf{I}$ and $\mathbf{J}_r$ are the body and the rotor inertia respectively. m is the mass of the system, b is the thrust factor, l is the distance from the center of mass to the rotor (it is assumed the same distance to all rotors) and $\boldsymbol{\tau}_a$ represents the torque applied to the quadrotor. For this configuration, $\boldsymbol{\tau}_a$ is denoted as follows

$$\boldsymbol{\tau}_a = \begin{pmatrix} \frac{\sqrt{2}}{2} lb\left(\Omega_1^2 + \Omega_2^2 - \Omega_3^2 - \Omega_4^2\right) \\ \frac{\sqrt{2}}{2} lb\left(\Omega_1^2 + \Omega_3^2 - \Omega_2^2 - \Omega_4^2\right) \\ d\left(\Omega_1^2 + \Omega_4^2 - \Omega_2^2 - \Omega_3^2\right) \end{pmatrix} \tag{3}$$

with a drag factor d.

The quadrotor dynamics, as described in [4], are given by

$$\begin{aligned}
\ddot{x} &= (\cos(\phi)\, sin(\theta)\sin(\psi) + \sin(\phi)\sin(\psi))\,\tfrac{U_1}{m} \\
\ddot{y} &= (\cos(\phi)\, sin(\theta)\sin(\psi) - \sin(\phi)\sin(\psi))\,\tfrac{U_1}{m} \\
\ddot{z} &= -g + (\cos(\phi)\cos(\theta))\,\tfrac{U_1}{m} \\
\ddot{\phi} &= \dot{\theta}\dot{\psi}\left(\tfrac{I_y - I_z}{I_x}\right) - \tfrac{J_r}{I_x}\dot{\theta}\Omega + \tfrac{l}{I_x}U_2 \\
\ddot{\theta} &= \dot{\phi}\dot{\psi}\left(\tfrac{I_z - I_x}{I_y}\right) + \tfrac{J_r}{I_y}\dot{\phi}\Omega + \tfrac{l}{I_y}U_3 \\
\ddot{\psi} &= \dot{\phi}\dot{\theta}\left(\tfrac{I_x - I_y}{I_z}\right) + \tfrac{U_4}{I_z}
\end{aligned} \tag{4}$$

where Ω represents the gyroscopic effects induced by the propellers, which are usually neglected [53], and they are given by

$$\Omega = \Omega_2 + \Omega_4 - \Omega_1 - \Omega_3 \tag{5}$$

U_i are the system inputs: U_1 is related to its translation and U_2 to U_4 are related to its attitude and orientation. The relation between both subsystems can be seen in Figure 4. The inputs U_i for this specific configuration of multirotor are given by

$$\begin{aligned}
U_1 &= b\left(\Omega_1^2 + \Omega_2^2 + \Omega_3^2 + \Omega_4^2\right) \\
U_2 &= \tfrac{\sqrt{2}}{2}b\left(\Omega_1^2 + \Omega_2^2 - \Omega_3^2 - \Omega_4^2\right) \\
U_3 &= \tfrac{\sqrt{2}}{2}b\left(\Omega_1^2 + \Omega_3^2 - \Omega_2^2 - \Omega_4^2\right) \\
U_4 &= b\left(\Omega_1^2 + \Omega_4^2 - \Omega_2^2 - \Omega_3^2\right)
\end{aligned} \tag{6}$$

and they are calculated by the neural network. In this work, a Multilayer Perceptron trained with the Extended Kalman Filter is selected.

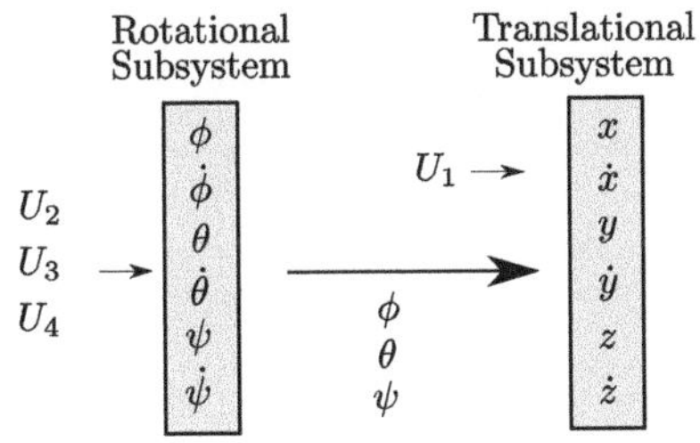

Figure 4. U_2, U_3 and U_4 are inputs for the rotational subsystem; U_1, roll, pitch and yaw are inputs for the following translation subsystem.

4. MLP trained with the EKF

The architecture of an ANN was inspired by biological neurons. Like synaptic connection in a regular biological neuron, each node in an ANN receives a signal from an adjacent node and its output is computed by some nonlinear function of the sum of the inputs. These connections between adjacent neurons have weights that adjusts as the network learns [19].

The training process of the MLP can be seen as an estimation problem for a nonlinear system that can be solved with the EKF [54–56]. Neural networks trained with the EKF have demonstrated faster learning speeds and convergence times than networks trained with algorithms based on backpropagation, which is ideal for real time applications [57,58]. The objective is to find the optimal weights that minimize the prediction error [59].

Consider an MLP with L weights and m output nodes. The neural network can be modeled as follows

$$\boldsymbol{w}(k+1) = \boldsymbol{w}(k) \tag{7}$$

$$\hat{\boldsymbol{y}}(k) = h\left(\boldsymbol{w}(k), \boldsymbol{u}(k)\right) \tag{8}$$

where $w(k)$ is the state vector, $u(k)$ is the input vector, $\hat{y}$ is the output vector and h is the nonlinear output function. From Kalman Filter equations, it is known that

$$\mathbf{K}(k) = \mathbf{P}(k)\mathbf{H}^T(k)\left[\mathbf{R}(k) + \mathbf{H}(k)\mathbf{P}(k)\mathbf{H}^T(k)\right]^{-1} \tag{9}$$

$$w(k+1) = w(k) + \eta\mathbf{K}(k)\left[y(k) - \hat{y}(k)\right] \tag{10}$$

$$\mathbf{P}(k) = \mathbf{P}(k) - \mathbf{K}(k)\mathbf{H}(k)\mathbf{P}(k) + \mathbf{Q}(k) \tag{11}$$

where L is the total number of weights, η is the learning rate, $\mathbf{P}(k) \in \Re^{L\times L}$ and $\mathbf{P}(k+1) \in \Re^{L\times L}$ are the prediction error covariance matrices in k and $k+1$. $\mathbf{K}(k) \in \Re^{L\times m}$ is the Kalman gain matrix. $y \in \Re^m$ is the system output, and $\hat{y}$ is the network output. $\mathbf{Q} \in \Re^{L\times L}$ is the process noise covariance matrix, and $\mathbf{R} \in \Re^{m\times m}$ represents the measurement covariance error. $w \in \Re^L$ is the weights (state) vector, and $\mathbf{H} \in \Re^{m\times L}$ contains the partial derivatives of each output of the neural network $\hat{y}$ with respect to each weight w_j and it is defined as

$$\mathbf{H}_{ij}(k) = \left[\frac{\partial \hat{y}_i(k)}{\partial w_j(k)}\right]_{w(k)=\hat{w}(k+1)}, i = 1\ldots m, j = 1\ldots L \tag{12}$$

Consider the MLP in Figure 5 with one hidden layer and one node at the output layer, where p denotes the number of inputs to the network and h the number of nodes in the hidden layer. The output of the neural network is defined by

$$\sigma_i = \frac{1}{1+e^{-n_i}} \quad i = 1\ldots h \tag{13}$$

$$n_i = \sum_{j=0}^{p} w_{ij}^{(1)} x_j \quad x_0 = +1 \tag{14}$$

$$v_1 = \sum_{k=0}^{h} w_{1k}^{(2)} u_k \quad u_0 = +1 \tag{15}$$

$$\hat{y} = v_1 \tag{16}$$

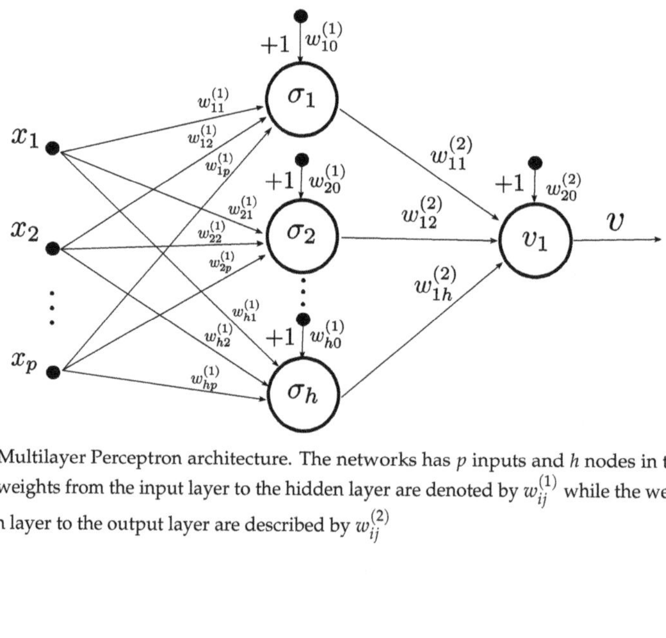

Figure 5. Multilayer Perceptron architecture. The networks has p inputs and h nodes in the hidden layer. the weights from the input layer to the hidden layer are denoted by $w_{ij}^{(1)}$ while the weights from the hidden layer to the output layer are described by $w_{ij}^{(2)}$

Finally, vector $\mathbf{H}$ can be expressed as

$$\mathbf{H} = \frac{\partial \hat{y}}{\partial w} = \left[\begin{array}{cccc} \frac{\partial \hat{y}}{\partial w_{10}^{(1)}} & \frac{\partial \hat{y}}{\partial w_{11}^{(1)}} & \cdots & \frac{\partial \hat{y}}{\partial w_{1h}^{(2)}} \end{array} \right] \tag{17}$$

where

$$\begin{array}{llll}
\frac{\partial \hat{y}}{\partial w_{10}^{(1)}} = \frac{w_{11}^{(2)} e^{-n_1}}{(1-e^{-n_1})^2} x_0; & \frac{\partial \hat{y}}{\partial w_{11}^{(1)}} = \frac{w_{11}^{(2)} e^{-n_1}}{(1-e^{-n_1})^2} x_1; & \cdots & \frac{\partial \hat{y}}{\partial w_{1p}^{(1)}} = \frac{w_{11}^{(2)} e^{-n_1}}{(1-e^{-n_1})^2} x_p; \\
\frac{\partial \hat{y}}{\partial w_{20}^{(1)}} = \frac{w_{12}^{(2)} e^{-n_2}}{(1-e^{-n_2})^2} x_0; & \frac{\partial \hat{y}}{\partial w_{21}^{(1)}} = \frac{w_{12}^{(2)} e^{-n_2}}{(1-e^{-n_2})^2} x_1; & \cdots & \frac{\partial \hat{y}}{\partial w_{2p}^{(1)}} = \frac{w_{12}^{(2)} e^{-n_2}}{(1-e^{-n_2})^2} x_p; \\
\vdots & \vdots & \vdots & \vdots \\
\frac{\partial \hat{y}}{\partial w_{h0}^{(1)}} = \frac{w_{1h}^{(2)} e^{-n_h}}{(1-e^{-n_h})^2} x_0; & \frac{\partial \hat{y}}{\partial w_{h1}^{(1)}} = \frac{w_{1h}^{(2)} e^{-n_h}}{(1-e^{-n_h})^2} x_1; & \cdots & \frac{\partial \hat{y}}{\partial w_{hp}^{(1)}} = \frac{w_{1h}^{(2)} e^{-n_h}}{(1-e^{-n_h})^2} x_p; \\
\frac{\partial \hat{y}}{\partial w_{10}^{(2)}} = 1; & \frac{\partial \hat{y}}{\partial w_{11}^{(2)}} = \frac{1}{1+e^{-n_1}}; & \cdots & \frac{\partial \hat{y}}{\partial w_{1h}^{(2)}} = \frac{1}{1+e^{-n_h}};
\end{array} \tag{18}$$

Let us define a new variable γ as follows

$$\gamma(n_i) = \frac{w_{1i}^{(2)} e^{-n_i}}{(1+e^{-n_i})^2}, \quad i = 1 \ldots h; \tag{19}$$

then, the vector H for the MLP shown in Figure 5 with sigmoid activation functions for the hidden layers and linear function for the output node can be expressed as follows

$$\mathbf{H} = \left[\begin{array}{ccccccccccc} \gamma(n_1) x_0 & \ldots & \gamma(n_1) x_p & \gamma(n_2) x_0 & \ldots & \gamma(n_h) x_p & u_0 & u_1 & \ldots & u_h \end{array} \right] \tag{20}$$

The network designed for this work has two inputs which are the error and the derivative of the error between the desired pose and the pose of the system which is calculated using monocular visual odometry.

5. Monocular Visual Odometry

Multirotors are equipped with inertial sensors which are capable of measuring the attitude and orientation of the system. Unfortunately, most of the quadrotors do not have any onboard sensors to measure position in indoor environments; in contrast, position is usually measured by a GPS sensor which has an error between 1 and 5 meters depending on the quality of the GPS signal.

This approach is based on Parallel Tracking and Mapping (PTAM) algorithm [60] to solve the localization problem. The algorithm is named this way because tracking and mapping are separated and run in parallel; it creates a keyframe based map using Bundle Adjustment (BA). This map is initialized from a stereo pair, and new points are initialized with epipolar searches.

It is well known that the scale of the environment cannot be determined using only monocular vision [52]. Once the map is initialized, the visual map is rotated such that the xy plane corresponds to the horizontal plane according to the accelerometer, and it is scaled with an average keypoint depth of 1. During the tracking, the scale of the map $\lambda \in \mathbb{R}$ must be estimated as in [52].

The quadrotor measures in regular intervals, the distance traveled according to the visual SLAM $\mathbf{x}_i \in \mathbb{R}^d$, and uses the metric sensors available, denoted by $\mathbf{y}_i \in \mathbb{R}^d$. To each interval, a pair $(\mathbf{x}_i, \mathbf{y}_i)$ is given, where $\mathbf{x}_i$ is scaled according to the visual map and $\mathbf{y}_i$ is in metric units. Both $\mathbf{x}_i$ and $\mathbf{y}_i$ measure motion of the UAV, and they are related by $\mathbf{x}_i \approx \lambda \mathbf{y}_i$. If Gaussian noise in the measurements is assumed, then a set of sample pairs $\{(\mathbf{x}_1, \mathbf{y}_1) \cdots (\mathbf{x}_n, \mathbf{y}_n)\}$ is given with

$$\mathbf{x}_i \sim \mathcal{N}(\lambda \mu_i, \sigma_x^2 \mathbf{I}_{dxd}) \tag{21}$$

$$\mathbf{y}_i \sim \mathcal{N}(\mu_i, \sigma_y^2 \mathbf{I}_{dxd}) \tag{22}$$

where $\mu_1 \cdots \mu_i \in \mathbb{R}^2$ are the unknown distances, $\sigma_x^2, \sigma_y^2 \in \mathbb{R}^+$ are the variances of the measurements error and I is the identity matrix of dimension d. To estimate the unknown parameters, maximum-likelihood estimation is used, minimizing the negative log-likelihood.

$$\mathcal{L}(\mu_1 \cdots \mu_n, \lambda) \propto \frac{1}{2} \sum_{i=1}^{n} \left(\frac{\|\mathbf{x}_i - \lambda \mu_i\|^2}{\sigma_x^2} + \frac{\|\mathbf{y}_i - \mu_i\|^2}{\sigma_y^2} \right) \tag{23}$$

The global minimum of (23) is unique; its derivation can be found in [52,61], and it is stated as follows

$$\mu_i^* = \frac{\lambda^* \sigma_y^2 \mathbf{x}_i + \sigma_x^2 \mathbf{y}_i}{\lambda^{*2} \sigma_y^2 + \sigma_x^2} \tag{24}$$

$$\lambda^* = \frac{s_{xx} - s_{yy} + sign(s_{xy}) \sqrt{(s_{xx} - s_{yy})^2 + 4s_{xy}^2}}{2\sigma_x^{-1} \sigma_y s_{xy}} \tag{25}$$

with

$$s_{xx} := \sigma_y^2 \sum_{i=1}^{n} \mathbf{x}_i^T \mathbf{x}_i \tag{26}$$

$$s_{yy} := \sigma_x^2 \sum_{i=1}^{n} \mathbf{y}_i^T \mathbf{y}_i \tag{27}$$

$$s_{xy} := \sigma_x \sigma_y \left(\sum_{i=1}^{n} \mathbf{x}_i^T \mathbf{y}_i \right)^2 \tag{28}$$

Assuming σ_x^2 and σ_y^2 are known, (25) gives a closed solution for the estimation of λ. For the estimation of the measurement variances, the authors refer to [52]. To generate the sample pairs, for each visual altitude measurement $a_v(t_i) \in \mathbb{R}$ there is a corresponding metric altitude measurement $a_m(t_i) \in \mathbb{R}$ over a window of sensor measurements, and then a visual and metric distance traveled within a period is computed as follows.

$$\mathbf{x}_i := a_v(t_i) - a_v(t_{i-k}) \tag{29}$$

$$\mathbf{y}_i := a_m(t_i) - a_m(t_{i-k}) \tag{30}$$

For visual odometry, EKF is used to identify and reject falsely tracked frames and compensate for time delays [6]. The filter includes the motion model of the quadrotor; the state space of the EKF is given by

$$\mathbf{x}_t := (x_t, y_t, z_t, \dot{x}_t, \dot{y}_t, \dot{z}_t, \Phi_t, \Theta_t, \Psi_t, \dot{\Psi}_t)^T \tag{31}$$

where (x, y, z) is the position of the quadrotor; $(\dot{x}, \dot{y}, \dot{z})$ its velocity; (Φ, Θ, Ψ) are the roll, pitch and yaw angle respectively (in degrees); and $\dot{\Psi}$ is the yaw rotational speed. For each sensor, an observation function is defined as below

$$h(\mathbf{x}_i) := (\dot{x}_t \cos\Psi_t - \dot{y} \sin\Psi_t, \dot{x}_t \sin\Psi_t + \dot{y} \cos\Psi_t, \dot{z}, \Phi_t, \Theta_t, \dot{\Psi}_t)^T \tag{32}$$

and the respective observation vector, derived from sensor measurements is defined as shown

$$\mathbf{z}_t := \left(\widehat{v}_{x,t}, \widehat{v}_{y,t}, \frac{\widehat{h}_t - \widehat{h}_{t-1}}{\delta_{t-1}}, \widehat{\Phi}_t, \widehat{\Theta}_t, \frac{\widehat{\Psi}_t - \widehat{\Psi}_{t-1}}{\delta_{t-1}} \right)^T \tag{33}$$

where $\widehat{v}_x, t$ and $\widehat{v}_y, t$ are velocities in xy plane directly measured by the quadrotor. The velocity in z direction can be computed since the altitude $\widehat{h}_t$ is given by an ultrasonic or an air pressure sensor and

δ_t is the time from time step t and $t+1$. Equally, for rotation readings, roll Φ and pitch Θ are directly measured and the yaw rotation speed can be calculated since Ψ is given by the IMU.

If PTAM tracks a video frame, the pose estimation is scaled with λ and transformed to the coordinate system of the quadrotor, leading to direct observation of its pose as follows.

$$h_p(\mathrm{x}_t) := (x_t, y_t, z_t, \Phi_t, \Theta_t, \Psi_t)^T \tag{34}$$

$$\mathrm{z}_p := f(E_{DC}E_{C,t}) \tag{35}$$

where $E_{C,t}$ is the estimated camera pose scaled with λ, E_{DC} is the rigid transformation from the camera to the quadrotor coordinate system and f is the function that maps from $SE(3)$ to roll-pitch-yaw representation.

The Kalman filter predicts how the state vector x_t evolves to the next time step. The horizontal acceleration is proportional to the horizontal force and is given by

$$\begin{pmatrix} \ddot{x} \\ \ddot{y} \end{pmatrix} \propto F_{acc} - F_{drag} \tag{36}$$

where F_{acc} is the drag force and F_{acc} denotes the accelerating force which is proportional to the projection of the z axis of the quadrotor onto the horizontal plane, which leads to

$$\ddot{x}(\mathrm{x}_t) = c_1(\cos\Psi_t \sin\Phi_t \cos\Theta_t - \sin\Psi_t \sin\Theta_t) - c_2\dot{x}_t \tag{37}$$

$$\ddot{y}(\mathrm{x}_t) = c_1(-\sin\Psi_t \sin\Phi_t \cos\Theta_t - \cos\Psi_t \sin\Theta_t) - c_2\dot{y}_t \tag{38}$$

where coefficients c_1 and c_2 are estimated from data collected from test flights. The influence of the control inputs $u_t = \left(\bar{\Phi}, \bar{\Theta}, \bar{\dot{z}}, \bar{\dot{\Psi}}\right)$ is described by

$$\dot{\Phi}(\mathrm{x}_t, \mathrm{u}_t) = c_3\bar{\Phi}_t - c_4\Phi_t \tag{39}$$

$$\dot{\Theta}(\mathrm{x}_t, \mathrm{u}_t) = c_3\bar{\Theta}_t - c_4\Theta_t \tag{40}$$

$$\ddot{\Psi}(\mathrm{x}_t, \mathrm{u}_t) = c_5\bar{\dot{\Psi}}_t - c_6\dot{\Psi}_t \tag{41}$$

$$\ddot{z}(\mathrm{x}_t, \mathrm{u}_t) = c_7\bar{\dot{z}}_t - c_8\dot{z}_t \tag{42}$$

where coefficients $(c_3, \cdots, c_8)$ are estimated from flight tests. The overall state transition is given by

$$\begin{pmatrix} x_{t+1} \\ y_{t+1} \\ z_{t+1} \\ \dot{x}_{t+1} \\ \dot{y}_{t+1} \\ \dot{z}_{t+1} \\ \Phi_{t+1} \\ \Theta_{t+1} \\ \Psi_{t+1} \\ \dot{\Psi}_{t+1} \end{pmatrix} \leftarrow \begin{pmatrix} x_t \\ y_t \\ z_t \\ \dot{x}_t \\ \dot{y}_t \\ \dot{z}_t \\ \Phi_t \\ \Theta_t \\ \Psi_t \\ \dot{\Psi}_t \end{pmatrix} + \delta_t \begin{pmatrix} \dot{x}_t \\ \dot{y}_t \\ \dot{z}_t \\ \ddot{x}(\mathrm{x}_t) \\ \ddot{y}(\mathrm{x}_t) \\ \ddot{z}(\mathrm{x}_t, \mathrm{u}_t) \\ \dot{\Phi}(\mathrm{x}_t, \mathrm{u}_t) \\ \dot{\Theta}(\mathrm{x}_t, \mathrm{u}_t) \\ \dot{\Psi}_t \\ \ddot{\Psi}(\mathrm{x}_t, \mathrm{u}_t) \end{pmatrix} \tag{43}$$

For a complete derivation of the state transition and delay compensation for a quickly reacting system to avoid oscillations and unstable behavior, the authors refer to [52].

6. Quadrotor Control Scheme

In this section, the control scheme is described. An MLP is used with an architecture as shown in Figure 5. The ANN has two inputs: the error between the desired value and the output measured from the system, and the derivative of this error to get information about the rate of change of the process output. PTAM estimates the positional observations (x, y, z), and the yaw angle ψ can be directly measured by the IMU.

The network has one hidden layer with four nodes and one neuron at the output layer, and there are four MLP modules, one for each controllable Degree of Freedom (Figure 6). The weights are randomly selected and uniformly distributed between $[-1, 1]$. The outputs of the four MLPs represent the control inputs U_i from (6). Despite all Degrees of Freedom being internally coupled, the advantage of using the MLP modules separately (one for every DoF) is an easier implementation and interpretation of the control scheme behavior.

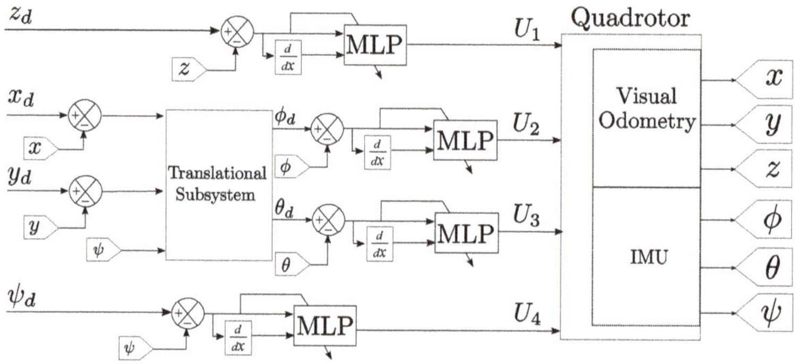

Figure 6. Control of quadrotor with MLP. There is one MLP module for each Degree of Freedom to control.

7. Results

The platform selected to test the algorithm was the Parrot Ar.Drone 2.0, which is a quadrotor equipped with a 3-axis gyroscope, 3-axis accelerometer, a magnetic compass, an ultrasound altimeter, one camera looking in x direction and another camera looking downward. For these experiments, the camera looking forward was used; it has a field of view of 92° and a resolution of 640×360. Its video is streamed at 30 fps to the ground station. Sensor measurements of the Ar.Drone 2.0 can be sent to the ground station at a frequency of 200 Hz; however, due to integration of the ANN training, control, localization and mapping, this transmission rate is decreased to 30 Hz in the following experiments.

7.1. Simulation Results

Simulation experiments were carried out in Linux Ubuntu with Gazebo, which is an open-source simulator with easy integration with a Robot Operating System (ROS). The world scene used for the test and the camera view are shown in Figure 7.

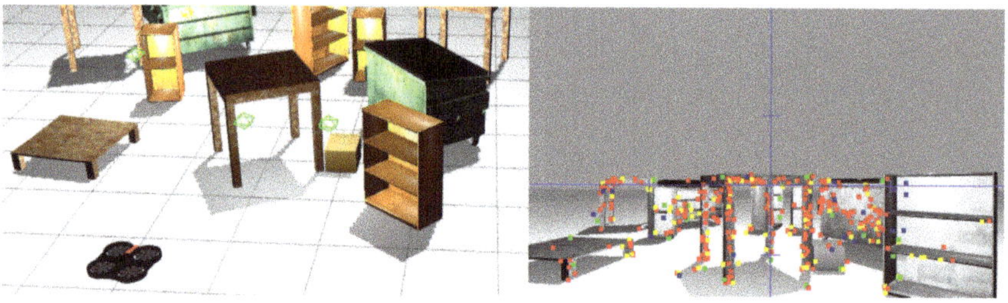

Figure 7. World Scene of Gazebo used for simulation and the camera view.

In the first experiment, a trajectory in the xy plane is selected to train weights for both directions at the same time. For each degree, all the weights of an MLP module with ten nodes in the hidden layer are selected. As can be seen in Figure 8, the overshoot is reduced and eventually eliminated after some iterations.

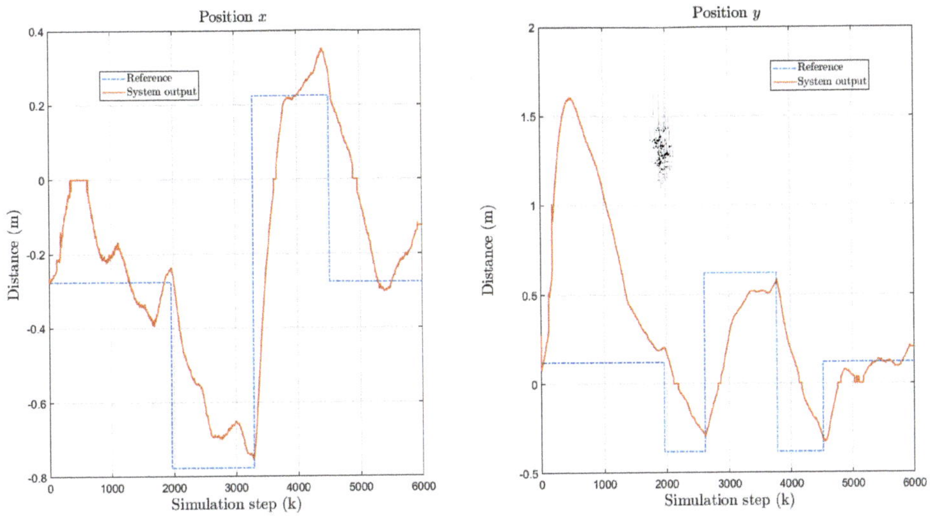

Figure 8. First simulation experiment with random initial weights. Dashed line represents the desired value and solid lines represent the system output. As can be seen, overshoot is reduced after some iterations.

After some iterations of x and y training, the z axis is added and a new test is performed training the three Degrees of Freedom. Again, the z axis MLP module is initialized with random weights. The results of these tests are shown in Figure 9. In Figure 10 the PTAM output can be seen. It can be seen that the overshoot in z is reduced after some iterations, and eventually, the path is followed correctly. Table 1 shows the Root Mean Squared Error (RMSE) for each axis.

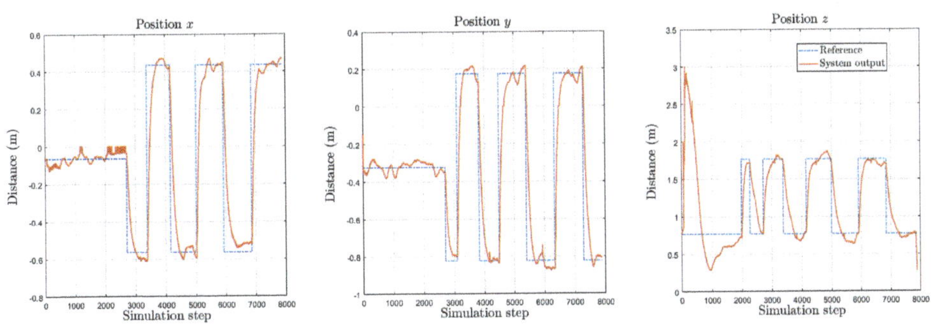

Figure 9. Second simulation: x and y axis were trained beforehand. Now z axis is added and initialized with random weights. Dashed line represents the reference and solid lines describe the output of the system.

Table 1. Root Mean Squared Error (RMSE) for each axis from simulation from Figure 10.

$RMSE_x$	$RMSE_y$	$RMSE_z$
0.1919	0.1989	0.2855

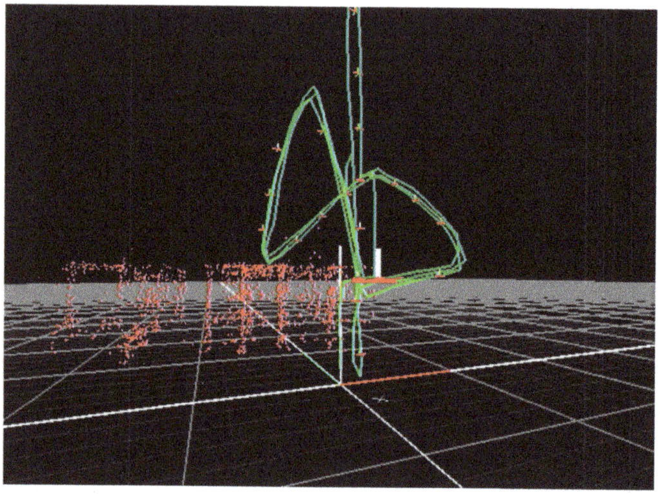

Figure 10. Parallel Tracking and Mapping (PTAM) output for the first trajectory. x, y and z axes are denoted red, green and white respectively. It can be seen that the quadrotor follows the reference correctly.

7.2. Experimental Results

The experimental tests were carried out in indoor unknown environment; however, since the localization of the system is estimated using only onboard sensors, it can also navigate in outdoor conditions. It is preferable that the scene includes sufficient objects (features) to be seen by the camera in a range of 0–7 m. The weights were initialized with the last value of the simulation. Since the Ar.Drone 2.0 driver for ROS was used, the parameters should be close to the real system. The data were received at 30 Hz (the time interval of each iteration was 1/30 s). Figure 11 shows the camera view and mapping output of the experiment. In Figure 12 are the results of controlling the quadrotor using the MLP using online training. As shown, oscillations and overshot were eliminated.

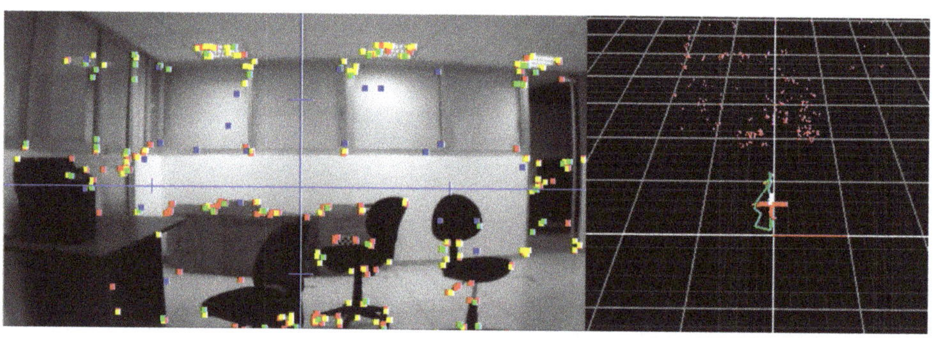

Figure 11. Camera view and map generated with PTAM for experimental tests.

For the yaw angle ψ, in contrast with the other three modules, the weights were initialized randomly. The fourth MLP module was trained online to control the yaw angle, and the results are shown in Figure 13. As can be seen, the neural network adapts its weights to follow the reference.

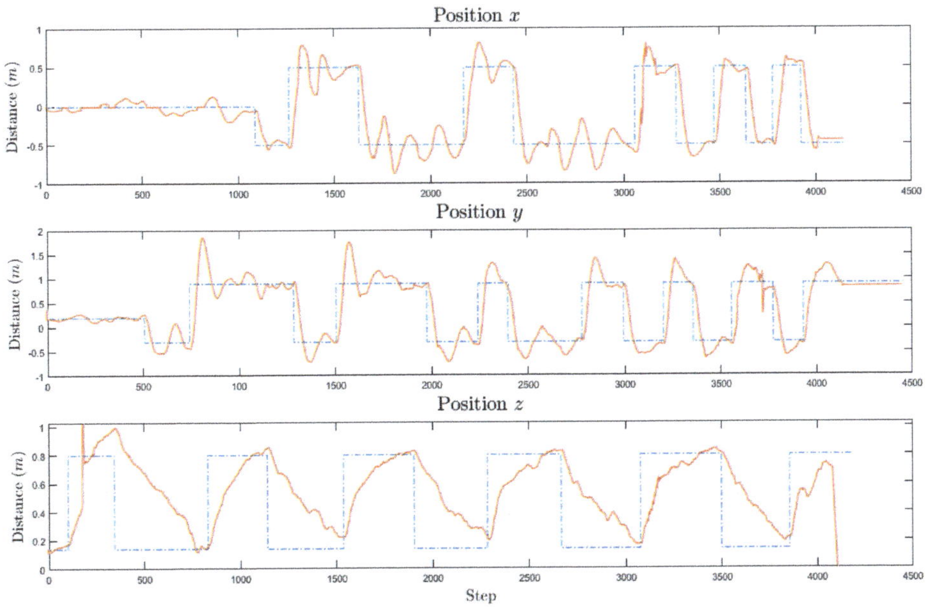

Figure 12. Experimental results for position of the UAV. One MLP module for each axis is used and they are trained online. The desired value is represented by the dashed line and the output of the system with the solid line.

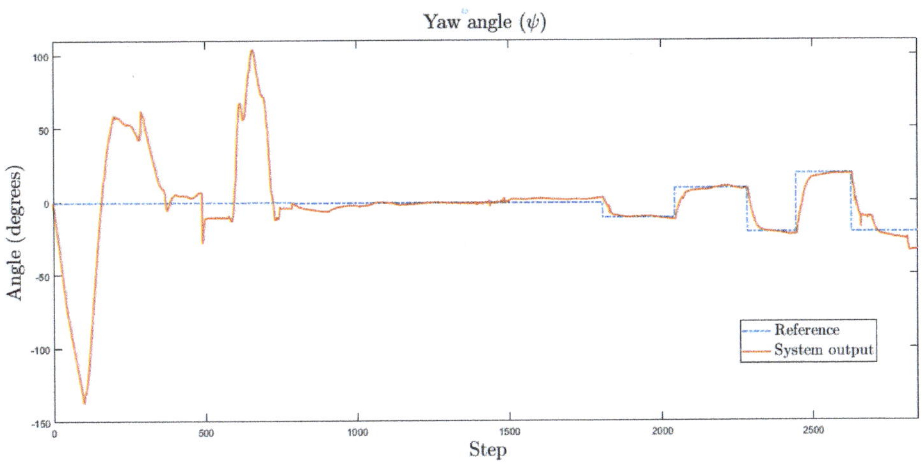

Figure 13. Experimental results for yaw angle of the UAV. Dashed line represents the reference and the solid line represents the system output.

For a second experiment, a mass is added to the system during the flight in order to test its adaptability; the mass has a value of 45 g, which represents, approximately, 1/3 of the maximum payload of the Ar.Drone 2.0. The results of this experiment are shown in Figure 14. As can be seen, at iteration 750 approximately, the mass is added on the x axis of the quadrotor. After some iterations, the oscillations are eliminated.

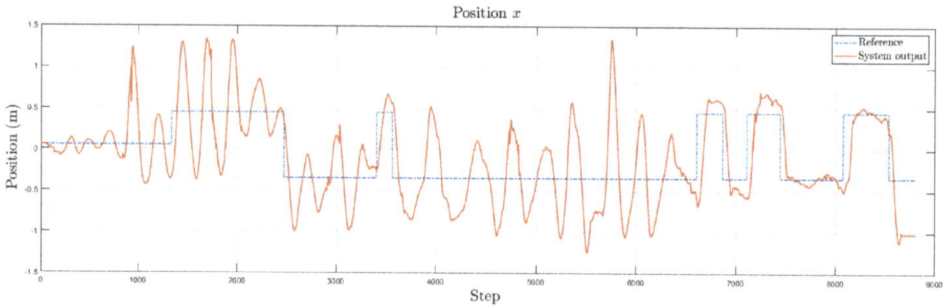

Figure 14. Experimental results for x axis changing the dynamics. Dashed line represents the reference and the solid line represents the system output.

8. Conclusions

In this paper, a direct control approach for a quadrotor was presented. The controller was based in a Multilayer Perceptron trained with the Extended Kalman Filter. Each MLP module consists of two inputs, ten nodes in the hidden layer and one output, which is the control action. Each controllable degree of freedom required an MLP module that was trained online to adapt its control law to uncertainties, loss of rotor efficiency, time delays and changes in the dynamic model of the system. The results were first validated via simulation using the Ar.Drone 2.0 ROS driver with Gazebo simulator. The neural network was initialized with random weights for x, y and z axes, and once the overshoot and oscillations were reduced, the weights were used to initialize the MLP modules in the real system. In the simulation, every degree of freedom was separately trained. For experimental tests, x, y, and z were initialized with the weights calculated in simulation, and all of them continued their training online. As can be seen in Figure 12, even using the simulation results, the neural networks must continue the training, since they present oscillations around the reference location. After some iteration steps, the oscillations were eliminated. For the experimental tests on the yaw angle (ψ), the weights were initialized randomly; that is, without previous training in a simulation. Figure 13 can be seen as the erratic behavior at the beginning of the experiment, and it is shown that the MLP adapt its weights to control the yaw of the quadrotor. Finally, in Figure 14, it has been shown that the controller adapts its weights when changing the dynamics during the flight.

Author Contributions: project administration, N.A.-D.; neural network formal analysis, A.Y.A.; neural network code, C.V.; ROS simulations and implementations, J.H.-B.; methodology, N.A.-D and C.L.-F.; real-time experiments, J.G.-A. and C.V.; writing—original draft preparation, J.G.-A. and C.V.; writing—review and editing, N.A.-D. and C.L.-F. All authors have read and agreed to the published version of the manuscript.

Funding: The authors thank the support of Council of Sciences and Technology (CONACYT), Mexico, for financially supporting the following projects: CB-256769, CB-258068 and PN-2016-4107.

Conflicts of Interest: The authors declare no conflict of interest.

Abbreviations

The following abbreviations are used in this manuscript:

ANN	Artificial Neural Network
BA	Bundle Adjustment
EKF	Extended Kalman Filter
GPS	Global Positioning System
HALE	High Altitude Long Endurance
LQR	Linear Quadratic Regulator
MALE	Medium Altitude Long Endurance
MLP	Multilayer Perceptron
PID	Proportional Integral Derivative

PTAM	Parallel Tracking and Mapping
ROS	Robot Operating System
SLAM	Simultaneous Localization and Mapping
UAV	Unmanned Aerial Vehicle
VTOL	Vertical Take-Off and Landing

References

1. Nebiker, S.; Annen, A.; Scherrer, M.; Oesch, D. A light-weight multispectral sensor for micro UAV—Opportunities for very high resolution airborne remote sensing. *Int. Arch. Photogramme. Remote Sens. Spat. Inf. Sci.* **2008**, *37*, 1193–1199.
2. Bento, M.d.F. Unmanned aerial vehicles: An overview. *Inside GNSS* **2008**, *3*, 54–61.
3. Eugster, H.; Nebiker, S. UAV-based augmented monitoring-real-time georeferencing and integration of video imagery with virtual globes. *IAPRSSIS* **2008**, *37*, 1229–1235.
4. Bouabdallah, S.; Murrieri, P.; Siegwart, R. Design and control of an indoor micro quadrotor. In Proceedings of the IEEE International Conference on Robotics and Automation, New Orleans, LA, USA, 26 April–1 May 2004; Volunm 5, pp. 4393–4398.
5. Bürkle, A.; Segor, F.; Kollmann, M. Towards autonomous micro uav swarms. *J. Intell. Robot. Syst.* **2011**, *61*, 339–353. [CrossRef]
6. Engel, J.; Sturm, J.; Cremers, D. Accurate figure flying with a quadrocopter using onboard visual and inertial sensing. *Imu* **2012**, *320(240)*. [CrossRef]
7. Rivera-Mejía, J.; Léon-Rubio, A.; Arzabala-Contreras, E. PID based on a single artificial neural network algorithm for intelligent sensors. *J. Appl. Res. Technol.* **2012**, *10*, 262–282. [CrossRef]
8. Lu, W.; Yang, J.; Liu, X. The PID Controller Based on the Artificial Neural Network and the Differential Evolution Algorithm. *JCP* **2012**, *7*, 2368–2375. [CrossRef]
9. Ge, S.S.; Zhang, J.; Lee, T.H. Adaptive neural network control for a class of MIMO nonlinear systems with disturbances in discrete-time. *IEEE Trans. Syst. Man Cybern. Part B Cybern.* **2004**, *34*, 1630–1645. [CrossRef]
10. Kalman, R.E. A new approach to linear filtering and prediction problems. *J. Basic Eng.* **1960**, *82*, 35–45. [CrossRef]
11. Sánchez Camperos, E.; Alanís García, A. *Redes Neuronales: Conceptos fundamentales y aplicaciones a control automático*; Pearson Prentice Hall: Upper Saddle River, NJ, USA, 2006.
12. Leung, H.; Haykin, S. The complex backpropagation algorithm. *IEEE Trans. Signal Process.* **1991**, *39*, 2101–2104. [CrossRef]
13. Ruck, D.W.; Rogers, S.K.; Kabrisky, M.; Maybeck, P.S.; Oxley, M.E. Comparative analysis of backpropagation and the extended Kalman filter for training multilayer perceptrons. *IEEE Trans. Pattern Anal. Mach. Intell.* **1992**, *14*, 686–691. [CrossRef]
14. De Freitas, J.F.; Niranjan, M.; Gee, A.H. Hierarchical Bayesian models for regularization in sequential learning. *Neural Comput.* **2000**, *12*, 933–953. [CrossRef] [PubMed]
15. Singhal, S.; Wu, L. Training feed-forward networks with the extended Kalman algorithm. In Proceedings of the International Conference on Acoustics, Speech, and Signal Processing, Glasgow, UK, 23–26 May 1989; pp. 1187–1190.
16. Williams, R.J. *Some Observations on the Use of the Extended Kalman Filter as A Recurrent Network Learning Algorithm*; Citeseer: University Park, PA, USA, 1992.
17. Latt, Z.K.K.; Si, H.; Kaneko, O. Controller Parameter Tuning of a Hexacopter with Fictitious Reference Iterative Tuning. In Proceedings of the 2019 SICE International Symposium on Control Systems (SICE ISCS), Kumamoto, Japan, 7–9 March 2019; pp. 96–101.
18. Serrano-Pérez, O.; Villarreal-Cervantes, M.G.; González-Robles, J.C.; Rodríguez-Molina, A. Meta-heuristic algorithms for the control tuning of omnidirectional mobile robots. *Eng. Optim.* **2019**, *52*, 325–342. [CrossRef]
19. Bari, S.; Hamdani, S.S.Z.; Khan, H.U.; ur Rehman, M.; Khan, H. Artificial Neural Network Based Self-Tuned PID Controller for Flight Control of Quadcopter. In Proceedings of the 2019 International Conference on Engineering and Emerging Technologies (ICEET), Lahore, Pakistan, 21–22 February 2019; pp. 1–5.

20. Radiansyah, S.; Kusrini, M.; Prasetyo, L. *Quadcopter Applications for Wildlife Monitoring*; IOP Conference Series: Earth and Environmental Science; IOP Publishing: Bristol, UK, 2017, Volume 54, p. 012066.
21. Antonio-Toledo, M.E.; Sanchez, E.N.; Alanis, A.Y.; Florez, J.; Perez-Cisneros, M.A. Real-time integral backstepping with sliding mode control for a quadrotor UAV. *IFAC-PapersOnLine* **2018**, *51*, 549–554. [CrossRef]
22. Wang, H.; Zhang, Y.; Yi, Y.; Xin, J.; Liu, D. Nonlinear tracking control methods applied to qball-x4 quadrotor uav against actuator faults. In Proceedings of the 2016 Chinese Control and Decision Conference (CCDC), Yinchuan, China, 28–30 May 2016; pp. 3478–3483.
23. Reyes-Valeria, E.; Enriquez-Caldera, R.; Camacho-Lara, S.; Guichard, J. LQR control for a quadrotor using unit quaternions: Modeling and simulation. In Proceedings of the CONIELECOMP 2013, 23rd International Conference on Electronics, Communications and Computing, Cholula, Mexico, 11–13 March 2013; pp. 172–178.
24. Cowling, I.D.; Whidborne, J.F.; Cooke, A.K. Optimal trajectory planning and LQR control for a quadrotor UAV. In the Proceedings of the International Conference on Control, Glasgow, UK, 2 September 2006.
25. Khatoon, S.; Gupta, D.; Das, L. PID & LQR control for a quadrotor: Modeling and simulation. In Proceedings of the 2014 International Conference on Advances in Computing, Communications and Informatics (ICACCI), New Delhi, India, 24–27 September 2014; pp. 796–802.
26. Lin, Q.; Cai, Z.; Wang, Y.; Yang, J.; Chen, L. Adaptive flight control design for quadrotor UAV based on dynamic inversion and neural networks. In Proceedings of the 2013 Third International Conference on Instrumentation, Measurement, Computer, Communication and Control, Shenyang, China, 21–23 September 2013; pp. 1461–1466.
27. Bouabdallah, S.; Siegwart, R. Full control of a quadrotor. In Proceedings of the 2007 IEEE/RSJ International Conference on Intelligent Robots and Systems, San Diego, CA, USA, 29 October–2 November 2007; pp. 153–158.
28. Madani, T.; Benallegue, A. Backstepping sliding mode control applied to a miniature quadrotor flying robot. In Proceedings of the IECON 2006—32nd Annual Conference on IEEE Industrial Electronics, Paris, France, 6–10 November 2006; pp. 700–705.
29. Swarnkar, S.; Kothari, M. A simplified adaptive backstepping control of aircraft lateral directional dynamics. *IFAC PapersOnLine* **2016**, *49*, 579–584. [CrossRef]
30. Meyer, G.; Su, R.; Hunt, L.R. Application of nonlinear transformations to automatic flight control. *Automatica* **1984**, *20*, 103–107. [CrossRef]
31. Lane, S.H.; Stengel, R.F. Flight control design using non-linear inverse dynamics. *Automatica* **1988**, *24*, 471–483. [CrossRef]
32. Ochi, Y.; Kanai, K. Design of restructurable flight control systems using feedback linearization. *J. Guid. Control Dyn.* **1991**, *14*, 903–911. [CrossRef]
33. Raffo, G.V.; Ortega, M.G.; Rubio, F.R. Backstepping/nonlinear Hinf control for path tracking of a quadrotor unmanned aerial vehicle. In Proceedings of the 2008 American Control Conference, Seattle, WA, USA, 11–13 June 2008; pp. 3356–3361.
34. Zhang, Y.; Chamseddine, A. Fault tolerant flight control techniques with application to a quadrotor UAV testbed. In *Automatic Flight Control Systems-Latest Developments*; IntechOpen: London, UK, 2012; pp. 119–150.
35. Li, T.; Zhang, Y.; Gordon, B.W. Nonlinear fault-tolerant control of a quadrotor uav based on sliding mode control technique. *IFAC Proc. Vol.* **2012**, *45*, 1317–1322. [CrossRef]
36. Dong, W.; Farrell, J.A.; Polycarpou, M.M.; Djapic, V.; Sharma, M. Command filtered adaptive backstepping. *IEEE Trans. Control Syst. Technol.* **2011**, *20*, 566–580. [CrossRef]
37. Guo, D.; Hu, H.; Yi, J. Neural network control for a semi-active vehicle suspension with a magnetorheological damper. *Modal Anal.* **2004**, *10*, 461–471. [CrossRef]
38. Lee, M.; Choi, H.S. A robust neural controller for underwater robot manipulators. *IEEE Trans. Neural Netw.* **2000**, *11*, 1465–1470. [PubMed]
39. Yuh, J. A neural net controller for underwater robotic vehicles. *IEEE J. Ocean. Eng.* **1990**, *15*, 161–166. [CrossRef]

40. Freddi, A.; Longhi, S.; Monteriù, A.; Prist, M. Actuator fault detection and isolation system for an hexacopter. In Proceedings of the 2014 IEEE/ASME 10th International Conference on Mechatronic and Embedded Systems and Applications (MESA), Senigallia, Italy, 10–12 September 2014; pp. 1–6.
41. Sheng, Q.; Xianyi, Z.; Changhong, W.; Gao, X.; Zilong, L. Design and implementation of an adaptive PID controller using single neuron learning algorithm. In Proceedings of the 4th World Congress on Intelligent Control and Automation, Shanghai, China, 10–14 June 2002; Volumn 3, pp. 2279–2283.
42. Grzonka, S.; Grisetti, G.; Burgard, W. Towards a navigation system for autonomous indoor flying. In Proceedings of the 2009 IEEE international conference on Robotics and Automation, Kobe, Japan, 12–17 May 2009; pp. 2878–2883.
43. Achtelik, M.; Bachrach, A.; He, R.; Prentice, S.; Roy, N. Stereo vision and laser odometry for autonomous helicopters in GPS-denied indoor environments. In *Unmanned Systems Technology XI*; International Society for Optics and Photonics: Bellingham, WA, USA, 2009; Volume 7332, p. 733219.
44. Strydom, R.; Thurrowgood, S.; Srinivasan, M.V. Visual odometry: Autonomous uav navigation using optic flow and stereo. In Proceedings of the Australasian Conference on Robotics and Automation, Melbourne, Australia, 2–4 December 2014.
45. Huang, A.S.; Bachrach, A.; Henry, P.; Krainin, M.; Maturana, D.; Fox, D.; Roy, N. Visual odometry and mapping for autonomous flight using an RGB-D camera. In *Robotics Research*; Springer: Berlin/Heidelberg, Germany, 2017; pp. 235–252.
46. Whelan, T.; Johannsson, H.; Kaess, M.; Leonard, J.J.; McDonald, J. Robust real-time visual odometry for dense RGB-D mapping. In Proceedings of the 2013 IEEE International Conference on Robotics and Automation, Karlsruhe, Germany, 6–10 May 2013.
47. Dryanovski, I.; Valenti, R.G.; Xiao, J. Fast visual odometry and mapping from RGB-D data. In Proceedings of the 2013 IEEE International Conference on Robotics and Automation, Karlsruhe, Germany, 6–10 May 2013; pp. 2305–2310.
48. Blösch, M.; Weiss, S.; Scaramuzza, D.; Siegwart, R. Vision based MAV navigation in unknown and unstructured environments. In Proceedings of the 2010 IEEE International Conference on Robotics and Automation, Anchorage, AK, USA, 3–7 May 2010; pp. 21–28.
49. Achtelik, M.; Achtelik, M.; Weiss, S.; Siegwart, R. Onboard IMU and monocular vision based control for MAVs in unknown in-and outdoor environments. In Proceedings of the 2011 IEEE International Conference on Robotics and Automation, Shanghai, China, 9–13 May 2011; pp. 3056–3063.
50. Forster, C.; Pizzoli, M.; Scaramuzza, D. SVO: Fast semi-direct monocular visual odometry. In Proceedings of the 2014 IEEE international conference on robotics and automation (ICRA), Hong Kong, China, 31 May–7 June 2014; pp. 15–22.
51. Engel, J.; Sturm, J.; Cremers, D. Camera-based navigation of a low-cost quadrocopter. In Proceedings of the 2012 IEEE/RSJ International Conference on Intelligent Robots and Systems, Vilamoura, Portugal, 7–12 October 2012; pp. 2815–2821.
52. Engel, J.; Sturm, J.; Cremers, D. Scale-aware navigation of a low-cost quadrocopter with a monocular camera. *Robot. Auton. Syst.* **2014**, *62*, 1646–1656. [CrossRef]
53. Schmidt, M.D. Simulation and Control of a Quadrotor Unmanned Aerial Vehicle. Master 's Thesis, University of Kentucky, Lexington, KY, USA, 2011.
54. Haykin, S. *Neural Networks: A Comprehensive Foundation*; Prentice Hall PTR: Upper Saddle River, NJ, USA, 1994.
55. Puskorius, G.V.; Feldkamp, L.A. Parameter-based Kalman filter training: Theory and implementation. *Kalman Filter. Neural Netw.* **2001**, 23. [CrossRef]
56. Williams, R.J. Training recurrent networks using the extended Kalman filter. In Proceedings of the IJCNN International Joint Conference on Neural Networks, Baltimore, MD, USA, 7–11 June 1992; Volume 4, pp. 241–246.
57. Haykin, S. *Kalman Filtering and Neural Networks*; John Wiley & Sons: Hoboken, NJ, USA, 2004; Volume 47.
58. Alanis, A.Y.; Sanchez, E.N. *Discrete-Time Neural Observers: Analysis and Applications*; Academic Press: Cambridge, MA, USA, 2017.
59. Camperos, E.N.S.; García, A.Y.A. *Redes neuronales: conceptos fundamentales y aplicaciones a control automático*; Pearson Educaci6n: London, UK, 2006.

60. Klein, G.; Murray, D. Parallel tracking and mapping for small AR workspaces. In Proceedings of the 2007 6th IEEE and ACM International Symposium on Mixed and Augmented Reality, Nara, Japan, 13–16 November 2007; pp. 1–10.
61. Engel, J. Autonomous Camera-Based Navigation of a Quadrocopter. Master's Thesis, TUM University, Munich, Germany, December 2011.

algorithms

MDPI

Article

Multi-Loop Model Reference Proportional Integral Derivative Controls: Design and Performance Evaluations

Baris Baykant Alagoz [1], Aleksei Tepljakov [2,*], Eduard Petlenkov [2] and Celaleddin Yeroglu [2]

1 Department of Computer Engineering, Inonu University, 44000 Malatya, Turkey; baykant.alagoz@inonu.edu.tr
2 Department of Computer Systems, Tallinn University of Technology, 12616 Tallinn, Estonia; eduard.petlenkov@taltech.ee (E.P.); c.yeroglu@inonu.edu.tr (C.Y.)
* Correspondence: aleksei.tepljakov@taltech.ee; Tel.: +90-555-273-70

Received: 27 December 2019; Accepted: 10 February 2020; Published: 13 February 2020

Abstract: Due to unpredictable and fluctuating conditions in real-world control system applications, disturbance rejection is a substantial factor in robust control performance. The inherent disturbance rejection capacity of classical closed loop control systems is limited, and an increase in disturbance rejection performance of single-loop control systems affects the set-point control performance. Multi-loop control structures, which involve model reference control loops, can enhance the inherent disturbance rejection capacity of classical control loops without degrading set-point control performance; while the classical closed Proportional Integral Derivative (PID) control loop deals with stability and set-point control, the additional model reference control loop performs disturbance rejection control. This adaptive disturbance rejection, which does not influence set-point control performance, is achieved by selecting reference models as transfer functions of real control systems. This study investigates six types of multi-loop model reference (ML-MR) control structures for PID control loops and presents straightforward design schemes to enhance the disturbance rejection control performance of existing PID control loops. For this purpose, linear and non-linear ML-MR control structures are introduced, and their control performance improvements and certain inherent drawbacks of these structures are discussed. Design examples demonstrate the benefits of the ML-MR control structures for disturbance rejection performance improvement of PID control loops without severely deteriorating their set-point performance.

Keywords: multi-loop model reference control; PID controllers; disturbance rejection control

1. Introduction

Control systems encounter unpredictable disturbances in real-world control applications. In order to maintain optimal control performance, real control systems should be designed to be robust enough against environmental disturbances, which are mainly unpredictable in character. From a practical point of view, performance degradations of classical control systems are largely caused by their dependence on predetermined system models. In the controller design phase, classical PID controllers are tuned to produce optimal control of a predetermined nominal plant model, which is indeed an idealized representation of a real-world system. The mathematical models of real systems can maintain their validity for predefined operating conditions and ranges, and only observable dynamics of the systems in these range and conditions can be represented by system models. Hence, they omit some inaccessible system dynamics and unknown disturbances, which are not rigorously investigated in the modeling stage. To enhance the representative capability of mathematical modeling, it is well-known that non-linear modeling [1], fractional-order modeling [2], and stochastic modeling [3] methods are often used. However, nowadays, optimal PID tuning methods still rely on integer order dynamics and linear system design approaches for the sake of simplicity and practical effectiveness. Therefore, the coverage potential of these modeling types to represent real-world conditions is quite limited, while mathematical models can represent the behavior of real systems in some presumed idealized conditions.

In general, optimal controller design in classical control has been addressed using integer order plant models and linear dynamics. Consequently, the designed controller law becomes mathematically optimal only for those ideal and linearized models operating in numerical simulation environments. However, when they are implemented in real-world conditions, practitioners commonly encounter the fact that the calculated controller coefficients are not practically optimal for real processes and fine-tuning efforts are needed for these controllers to achieve acceptable control performance. Firstly, non-ideal realization of controller function inherently deteriorates control optimality. Secondly, a number of factors such as unmodelled dynamics, noise, environmental disturbance, ageing of system components, and parametric perturbations may have a significant influence on control performance. Therefore, it is obvious that model-based optimal controller design methods should adhere to robust control performance requirements in the controller design stage, including considerations of disturbance rejection performance and insensitivity to model parameter variation. The current study focuses on enhancement of the disturbance rejection performance of classical control loops. The disturbance model is assumed to be additive input disturbance that has unknown characteristics. The disturbance rejection capacity of a classical negative feedback PID control loop is inherently bound [4]. A feasible solution to increase disturbance rejection performance beyond the inherent bounds of a PID control loop is by using a multi-loop control structure [5]. The findings of the current study suggest that a multi-loop model reference (ML-MR) adaptive control structure could be an effective solution to improve the disturbance rejection performance of classical PID control loops.

Due to a performance tradeoff between disturbance rejection control and set-point control performances [4], increasing the disturbance rejection control performance can severely deteriorate the set-point performance of classical PID control loops. A primary motivation of this study is the improvement of disturbance rejection control performance without effecting the set-point performance of existing closed loop PID control systems.

ML-MR control structures are commonly composed of two hieratical loops. These are the reference model loop, which describes the desired response of a system, and the control loop, which deals with the stability of control systems. The control loop is the primary loop that maintains the stability of the system. The secondary loop is the reference model loop, also called the adaptation loop, which handles the preservation of the desirable control performance of the system, which is described by the reference model response. When a disturbance acts on the control loop, the adaptation loop contributes to the control efforts for the reduction of the negative effects of disturbances on system outputs, thus restoring control performance by means of reference responses that are generated by the reference model.

In literature, numerous ML-MR control structures have been proposed and their improvements in terms of control performance have been demonstrated. In former works, direct model reference adaptive control (MRAC) structures have been addressed [5–10]. Direct MRAC structures perform online self-tuning of controller parameters according to a predefined reference model [5–8]. Updating controller parameters in control action, known as online controller tuning, may lead to short-term instability and control performance degradations. Therefore, the stability and robust control performance of MRAC structures in real conditions remain questionable and direct MRAC control methodologies have not been accepted or adopted by practitioners [8]. The industrial applications of the MRAC method have remained limited in comparison to the widespread adoption of classical PID control loops. PID control loops have been used as industrial standard control systems because of their well-established design scheme. The simple structure and reliability of PID control law, which facilitates realization in real systems, are prominent in industrial control applications. Barkana has addressed performance and stability issues of direct MRAC methods and presented a stable direct MRAC methodology according to passivity conditions [8]. However, to avoid online controller parameter updates in control action, one branch of ML-MR control research studies aims to combine a classical PID control system with reference model control in a hierarchical multi-loop manner in order to benefit from the stability and set-point performance of the classical control loop and the robust control performance improvements of the model reference control (MRC) system.

The topic of multi-loop control involves a wide variety of control systems that employ multi-loop control structures, including cascaded control loops [11,12], multi-input multi-output control systems [13], use of additional control loops for performance improvements [14,15], direct model reference control based on output matching [16,17], direct model reference adaptive control with online controller tuning [5–10], model reference adaptive control based on Massachusetts Institute of Technology (MIT) rule [18–23], backstepping-based adaptive PID control [24], and hierarchical ML-MR adaptive control [25–28]. The current study investigates several configurations of hierarchical ML-MR adaptive control structures.

Hierarchical ML-MR adaptive PID control has been investigated, and certain benefits of this structure related to fault tolerance [25] and disturbance rejection [26] have been discussed. This structure can be easily applied to the existing closed loop PID control systems, thus transforming the PID control systems into model reference adaptive PID control systems. This presents a significant potential to improve the robust control performance of industrial PID control systems without largely modifying the existing PID loops. Although a growing research trend has been initiated for the use of ML-MR adaptive PID structures to improve the control performance of the closed loop PID control family [25–27], this control structure is still in the early stages of development. There is a need for further research efforts to establish design methodologies and reveal some practical benefits and drawbacks. Therefore, this study aims to investigate the benefits and drawbacks of ML-MR PID-MIT control [25,28] and addresses design and performance improvement problems in ML-MR structures. To this end, this paper revisits recent multi-loop model reference adaptive PID control techniques, and analytical tuning and performance issues are considered. In analytical optimal tuning of PID controllers, we adopt the integral time absolute error (ITAE) design rule that was proposed by Tavakoli et al. for control of the first order plus dead time dynamic systems. This design rule is based on dimensional analysis, which has been widely utilized to solve complex high-dimensional problems by reducing the number of variables to an essential set [29]. The main reason to use the Tavakoli–Tavakoli ITAE design rule for optimal tuning of PID control loops is that the method limits overshoot with faster establishment of set-point control in time delay systems [29].

The organization of this paper is given as follows. Section 2 details the mathematical background of the multi-loop model reference PID-MIT (ML-MR PID-MIT) control, and presents analytical equations for theoretical background and analytical tuning. After discussing some inherent drawbacks of ML-MR PID-MIT structures (e.g., nonlinearity and amplitude dependence of disturbance rejection performance), several ML-MR control structures are investigated to resolve these drawbacks in the following sections. In these sections, two types of hierarchical ML-MR control structures are considered.

(i) Reference modeling of closed loop control loops: These types of MRC structures use a reference model to describe the desired response of closed loop control systems. Section 3 presents a multi-loop model reference PID-MIT control with controller gain modification (ML-MR PID-MIT-CGM) that allows dynamic disturbance rejection via adaptively adjusting Reference to Disturbance Ratio (RDR) performance. To deal with the drawbacks of nonlinearity from the MIT rule, a multi-loop model reference PID integral (ML-MR PID-I) control, which performs a linear adaptation rule, is introduced in this section. Section 4 introduces a multi-loop model reference PID-IM control structure (ML-MR PID-IM) that is proposed as an internal model linear adaptation rule to implement an internal model control (IMC) approach.

(ii) Reference modeling of plants: These types of MRC structures use a reference model to describe desired responses of a process or a plant function. Section 5 presents the multi-loop model reference MIT-PID control, which employs the model reference control rule only for plant function adaptation (ML-MR MIT-PID-PFA) in contrast to ML-MR MIT-PID in Section 2. The main advantages of this type of structure come from the fact that all adaptation efforts aim for approximation of plant or process models compared to the reference model. This property is advantageous for time delay systems. Section 6 presents a multi-loop model reference PID-PID control with plant function adaptation (ML-MR PID-PID-PFA). This structure employs a linear adaptation rule for plant function and it can improve control performance for time delay systems.

Illustrative design examples cover control the problems of an automatic voltage regulator model, the liquid level of reboilers, large time delay experimental process models, and Twin-Rotor Multi-input multi-output System (TRMS) experimental setup.

2. Multi-Loop Model Reference PID-MIT (ML-MR PID-MIT) Control: Analytical Tuning and Nonlinearity

The ML-MR PID-MIT control structure has been proposed to enhance the robust control performance of existing closed loop control systems. Its application has been discussed in several works [25–28]. This structure is essentially composed of two hierarchical control loops: the inner loop is a classical PID control loop that is a stable and well-tuned control loop, while the outer loop is a adaptation loop that performs the MIT rule. The outer loop is appended to the inner loop to enhance the robust control performance of the inner loop according to regulation of a reference model. Figure 1 shows a block diagram of the ML-MR PID-MIT control structure. It has two error signals per loop:

(i) Control error: When the controller of the inner loop is tuned properly, the value of the control error, which is written as $e_c = u_r - y$, moves to zero, and thus ensures the control system output settles to the reference input (the signal u_r is the modified reference input controlled by the MIT rule and y is the system output).
(ii) Model error: Discrepancy between the reference model output and controlled system output is defined as the model error, which is written as $e_m = y - y_m$ (the signal y_m is the output of reference model). The objective of the outer loop is to shape the reference input signal (r) such that the model error converges to zero. The convergence of the model error to zero implies that the inner loop resembles the response of the reference model $T_m(s)$. The reference model describes the desired response of the control system that forms the inner loop.

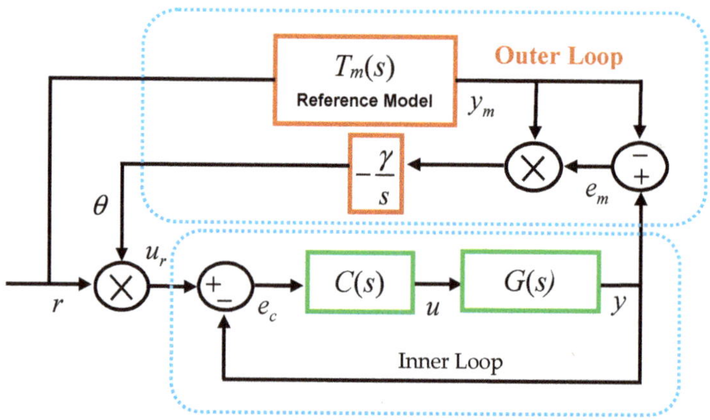

Figure 1. Block diagram of multi-loop model reference PID-MIT (ML-MR PID-MIT) control structure with input shaping.

Let us consider a PID control system as the inner control loop. The PID controllers are three-coefficient standard industrial controllers and optimal tuning of the PID coefficients (k_p, k_i,

k_d) enhances step performance by reducing overshoots and settling time. The transfer function of PID controllers is widely expressed in the parallel form as:

$$C(s) = k_p + \frac{k_i}{s} + k_d s \tag{1}$$

For a given plant (process) function $G(s)$, the transfer function of the inner loop can be written by:

$$T(s) = \frac{C(s)G(s)}{1 + C(s)G(s)} \tag{2}$$

Previous studies on the ML-MR PID-MIT control structure aim to maintain the initial well-tuned control performance of an existing closed loop control system. For this reason, the reference model is taken as the transfer function of the inner loop; that is, $T_m(s) = T(s)$. Accordingly, any performance deterioration in the inner loop leads to an increase in model errors. The recovery of temporal performance deterioration is carried out by shaping the reference input of the inner loop via $u_r = \theta r$. The adaptation gain θ is determined according to the well-known MIT rule [21], which implements continuous gradient descent optimization of the cost function:

$$J = \frac{1}{2}{e_m}^2 \tag{3}$$

The feed-forward MIT rule is expressed in the continuous domain as [18–21]:

$$\frac{d\theta}{dt} = -\gamma \frac{dJ}{d\theta}, \tag{4}$$

where the parameter γ is the learning rate, which is an important parameter for the convergence speed and stability of the gradient descent optimization. Taking Laplace transform of Equation (4), one obtains:

$$\theta = -\gamma \frac{1}{s} \frac{dJ}{d\theta} = -\gamma \frac{1}{s} e_m \frac{de_m}{d\theta} \tag{5}$$

To find the sensitivity derivative $\frac{de_m}{d\theta}$, the model error is written in the form of $e_m = y - y_m = T(s)\theta r - T_m(s) r$. Then, the sensitivity derivative of the system is obtained:

$$\frac{de_m}{d\theta} = T(s) r \tag{6}$$

Here, one can substitute the reference input with $r = y_m / T_m(s)$ in Equation (6) and rearrange the sensitivity derivative as:

$$\frac{de_m}{d\theta} = \frac{T(s)}{T_m(s)} y_m \tag{7}$$

By using it in Equation (5), the MIT rule for the update of the adaptation coefficient θ to minimize cost function J is obtained as [25,26,28]:

$$\theta = -\gamma \frac{1}{s} \left(\frac{T(s)}{T_m(s)} y_m e_m \right) \tag{8}$$

When one configures $T_m(s) = T(s)$, the adaptation gain θ is simplified to:

$$\theta = -\gamma \frac{1}{s} y_m e_m, \tag{9}$$

which is implemented in the outer loop in Figure 1 [25,26,28].

Application of ML-MR PID-MIT control for the existing control loop was defined as follows [25,26,28]:

Step 1: Identify the transfer function model of the existing closed loop PID control system by means of closed loop model identification methods and use this transfer function as a reference model.

Step 2: Enclose the existing closed loop control system via the outer loop. The outer loop employs Equation (9) and the reference model control.

To create a design from scratch for a first order system model in the form of $G(s) = \frac{K}{\tau s+1}e^{-Ls}$, an analytical tuning scheme for the ML-MR PID-MIT control structure can be proposed as follows:

Step 1: Design a closed loop PID control system according to the Tavakoli–Tavakoli PID tuning rule that can be rearranged for a standard PID controller function in parallel form as follows [29]:

$$k_p = (1/K)\frac{0.8}{(L/\tau)+0.1},\ k_i = k_p\frac{1}{(0.3L+\tau)},\ k_d = k_p\frac{0.06L}{(L/\tau)+0.04} \tag{10}$$

This optimal tuning rule implements ITAE and its control performance have been shown previously [29]. The optimal PID controller function of the inner loop can be written for the Tavakoli-Tavakoli optimal PID tuning rule as:

$$C(s) = (1/K)\frac{0.8}{(L/\tau)+0.1}\left(1+\frac{1}{(0.3L+\tau)s}+\frac{0.06L}{(L/\tau)+0.04}s\right) \tag{11}$$

Step 2: The reference model is a theoretical model that represents the optimally tuned closed loop control system. Therefore, the transfer function of reference model can be expressed as:

$$T_m(s) = \frac{C(s)G(s)}{1+C(s)G(s)} \tag{12}$$

In MATLAB and Simulink control system simulations, we implemented this reference model as a numerical model of the PID control loop, as illustrated in the diagram in Figure 2.

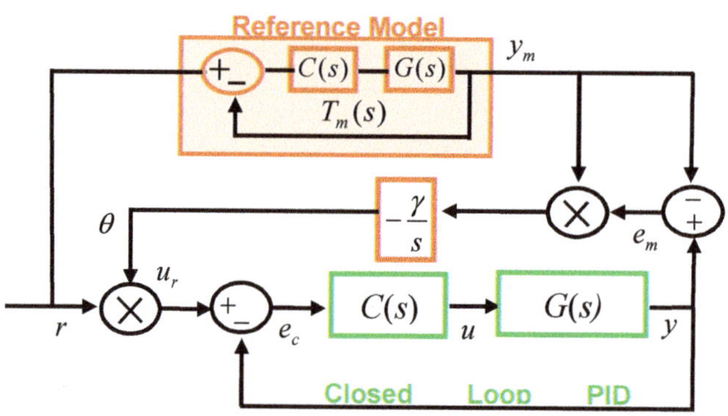

Figure 2. Block diagram of the ML-MR PID-MIT control structure and a solution for numerical realization of reference models.

Step 3: Implement the ML-MR PID-MIT control structure as shown in Figure 2. The learning rate $\gamma = |k_i|$ can be selected to comply with integral dynamics of the PID control loop.

It is important to note two characteristic features of this structure. Firstly, the performance of the ML-MR PID-MIT control structure is sensitive to the time delay component of the system. As the time delay of the system increases, the adaptation performance of the MIT rule decreases because large delay in the system response easily misleads gradient descent optimization of the MIT rule. Figure 3 shows this effect for the ML-MR PID-MIT control structure design for control of the plant functions with zero time delay ($G_1(s) = \frac{2}{s+1}$) and 0.3 s time delay ($G_2(s) = \frac{2}{s+1}e^{-0.3s}$) [29]. The system responses in the figure demonstrate that the ML-MR PID-MIT control structure can significantly improve the disturbance rejection performance of classical PID control for zero delay systems. However, when the time delay of the plant increases, improvement in disturbance rejection control decreases.

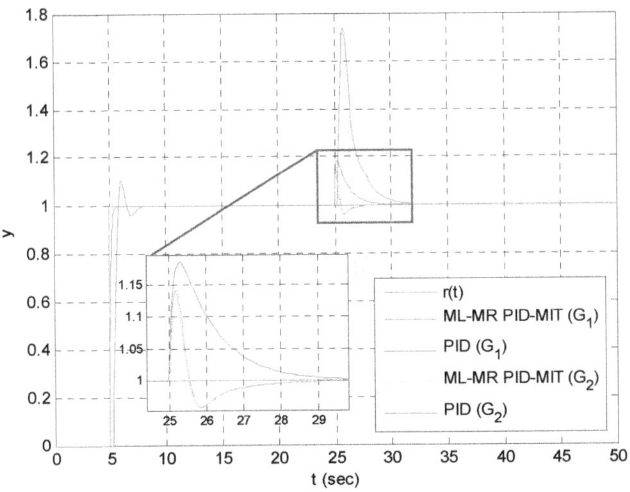

Figure 3. Comparison of disturbance rejection performance of ML-MR PID-MIT control and classical PID control for $G_1(s)$ and $G_2(s)$ plants.

Secondly, due to multiplication terms $u_r = \theta r$ and $y_m e_m$, the ML-MR PID-MIT control structure presents a nonlinear control characteristic. The origin of the nonlinearity in the adaptation gain θ is the nonlinear cost function, which involves the square of model error in Equation (3). This yields a nonlinear term $y_m e_m = y_m(y - y_m) = y_m y - y_m^2$. Therefore, ML-MR PID-MIT control can exhibit some characteristic properties that do not appear in linear control systems. A noteworthy property is that the time response of the PID-MIT control shows amplitude dependence. In the case of reference input $r(t) = 0$, this results in $y_m \cong 0$; in this case, the outer loop is not functional, so that $y_m e_m = 0$ in Equation (9). As the amplitude of the reference input signal $r(t)$ grows, the amplitude of y_m increases, and this effect indeed amplifies the model error due to the term of $y_m e_m$. Accordingly, this makes the MIT rule of the outer loop more sensitive to model errors and the system responds more effectively in cases where the model mismatches between reference model and control loops. To demonstrate this effect, an illustrative example for control of the plant function $G(s) = \frac{2}{s+1}$ is carried out.

Figure 4 shows time response ML-MR PID-MIT control structure that is designed for plant function $G(s) = \frac{2}{s+1}$ [29]. The learning rate γ is set to -4, and the PID controller is configured to $k_p = 4$, $k_i = 4$, and $k_d = 0$ according to the Tavakoli–Tavakoli PID tuning rule in this simulation. When the amplitude of $r(t)$ moves to one ($r(t) = 1$), the ML-MR PID-MIT control structure becomes more effective and contributes to the disturbance rejection control performance of the inner loop (classical PID control loop). However, when the amplitude of $r(t)$ moves to zero ($r(t) = 0$), the ML-MR PID-MIT control structure is not functional, and its disturbance rejection control performance becomes the same as the disturbance rejection performance of the classical PID controller. The overlapping of time responses at set-point zero in the figure apparently confirms this effect.

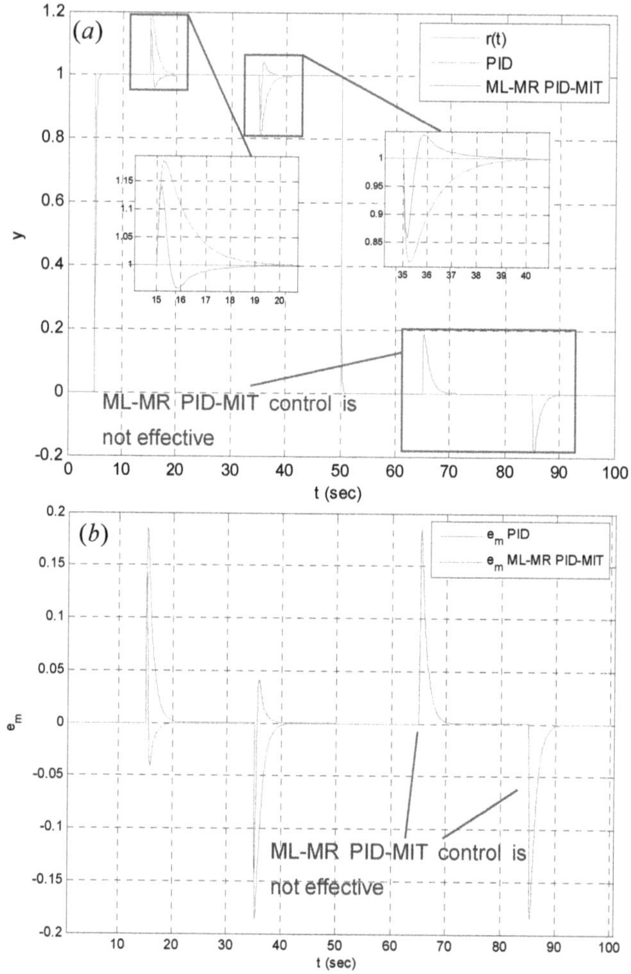

Figure 4. Comparison of disturbance rejection performances of ML-MR PID-MIT control and classical PID control for the cases of $r(t) = 1$ and $r(t) = 0$. (**a**) Step response and (**b**) change of model error.

3. Multi-Loop Model Reference PID-MIT with Controller Gain Modification (ML-MR PID-MIT-CGM): Adaptive RDR Adjustment for Dynamic Disturbance Rejection and the Linear Adaptation Rule

The disturbance rejection capacity of a closed loop control system can be expressed with the reference to disturbance ratio (RDR) spectrum, which presents a measure for the additive input disturbance rejection performance of a control system for each frequency component [4,30]. The analysis demonstrated that the RDR performance of a closed loop control system depends on the spectral power density of the controller function [4]. The RDR spectrum of the PID controller is expressed in the form of [4]:

$$RDR_{PID}(\omega) = k_p^2 + (k_d\omega - \frac{k_i}{\omega})^2 \quad (13)$$

The increase of low frequency RDR values improves the disturbance rejection control for low frequency and steady disturbance signals. An increase in high frequency RDR values improves the transient disturbance rejection performance. However, as mentioned previously, the tradeoff between disturbance rejection control and set-point control performance bounds the RDR performance of the closed loop PID control systems [4]. A large increase in RDR values deteriorates the step response performance; this results in very high overshoots and consecutive ripples, which delay the convergence of the controlled system output to the set-point. Further increases in RDR values finally lead to instability of the control systems [4]. Therefore, the unstable state of the closed loop control systems limits the improvement of their RDR performances. A solution for this problem is adaptive adjustment of RDR performance; RDR performance should be increased for disturbance incidents in the inner loop. In order to adjust the RDR performance of the PID control loop adaptively, a variable gain PID controller is proposed by multiplying the PID controller function with the adaptation coefficient θ. This modified PID controller function allows adjustment of the DC gain of the classical PID controller by means of the coefficient θ. The transfer function of this controller can be expressed as:

$$C_{\theta}(s) = \theta(k_p + \frac{k_i}{s} + k_d s) \tag{14}$$

The RDR spectrum of this RDR-adjustable PID controller can be obtained as:

$$RDR_{PID\theta}(\omega) = \theta^2(k_p^2 + (k_d\omega - \frac{k_i}{\omega})^2) \tag{15}$$

This expression of the RDR spectrum apparently shows that the proposed variable gain PID controller $C_{\theta}(s)$ allows modification of the RDR spectrum via adaptation gain θ. This enables adaptive adjustment of the disturbance rejection performance of the inner loop by the outer loop. Since the gain θ amplifies the control error inside the inner loop, the value of the adaptation gain θ should be configured to remain positive, and hence is written by:

$$\theta = \left|\gamma \frac{1}{s} y_m e_m\right| \tag{16}$$

Figure 5 shows the RDR spectrum of the controller $C_{\theta}(s)$ for several values of the gain θ. This figure reveals that the proposed variable gain PID controller $C_{\theta}(s)$ enables modification of the RDR spectrum via the gain θ, and this asset allows adaptive adjustment of disturbance rejection performance in the control loop. For $\theta = 1$, the $C_{\theta}(s)$ controller exhibits disturbance rejection performance equal to the performance of the classical PID controllers. When $\theta > 1$, the disturbance rejection performance of $C_{\theta}(s)$ increases, as illustrated in the RDR spectrum in Figure 5. It should be noted that the variable gain PID control may cause instability, particularly for time delay systems, because higher values of the gain θ result in greater amplification of the integral gain, which may cause instability in the control systems as a consequence of accumulation of integral element control errors during the delay period of the system response. This limits the application domain of the proposed method to stable and almost-zero time delay plant functions.

The application steps for ML-MR PID-MIT with controller gain modification (ML-MR PID-MIT-CGM) for existing control loops can be given as follows:

Step 1: Obtain a transfer function model of an optimal control loop and use this transfer function as a reference model of the MRC loop ($T_m(s) = T_1(s)$).

Step 2: Enclose the existing closed loop control system by the outer loop, as shown in Figure 6. Apply the MIT rule for controller gain modification of the closed loop PID control system to adjust the RDR.

To show an application of the ML-MR PID-MIT-CGM control structure, an Automatic Voltage Regulator (AVR) control example is presented in this section. Figure 7 shows a block diagram of the proposed ML-MR PID-MIT-CGM control structure for the AVR control example. The learning rate γ is set to 10.

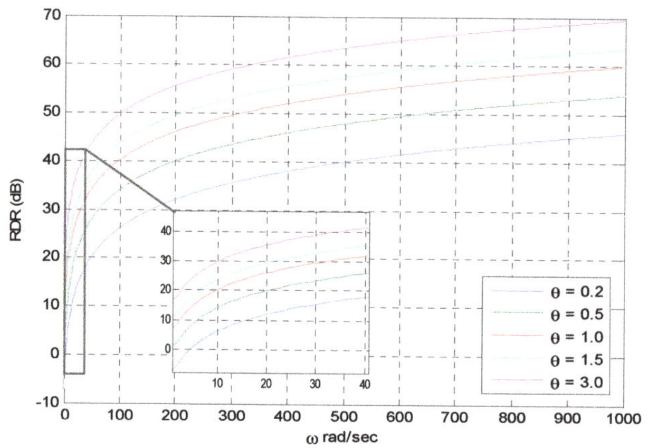

Figure 5. RDR spectra of $C_{\theta}(s)$ controllers for several θ values. This figure reveals the adjustment of RDR performance according to several θ configurations.

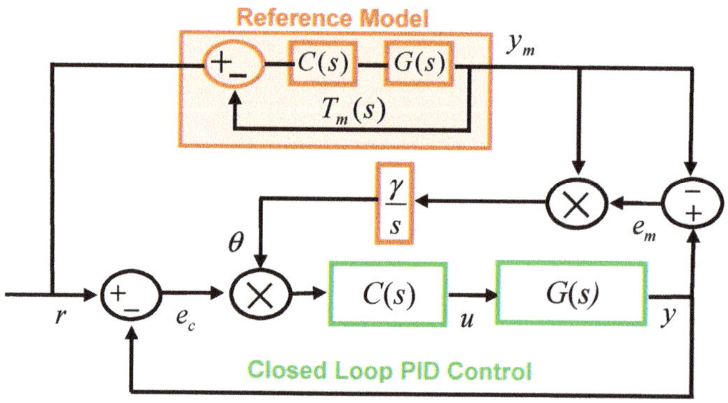

Figure 6. Block diagram of the ML-MR PID-MIT with controller gain modification (ML-MR PID-MIT-CGM) control structure and numerical realization of the reference model.

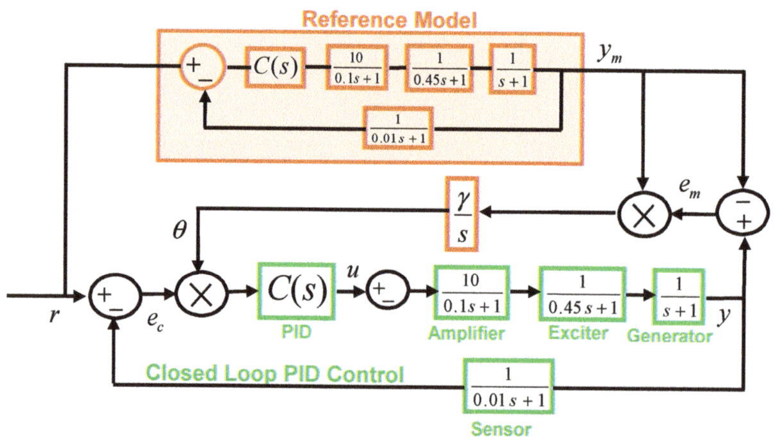

Figure 7. ML-MR PID-MIT-CGM control structure with control gain shaping that is implemented for improvement of the disturbance rejection performance of an optimal closed loop PID control of an AVR model.

Bendjeghaba et al. implemented an improved harmony search algorithm for optimal tuning of closed loop PID control of an AVR model. The optimal PID controller was obtained by Bendjeghaba et al. in [31] as:

$$C(s) = 0.6739 + \frac{0.5076}{s} + 0.2699s \tag{17}$$

We used this optimal PID controller for the AVR model in Figure 7. Figure 8a,b show responses of Bendjeghaba et al.'s optimal PID control loop and ML-MR PID-MIT-CGM control structure for square waveform input disturbances. The figure clearly shows disturbance rejection performance improvements for the ML-MR PID-MIT-CGM control at simulation times of 15 and 35 s, where the set-point is equal to 1. At these instances, the adaptation gain θ of the outer loop actively responds to step-up and step-down disturbances, and it increases the RDR performance of the inner loop by increasing θ, as illustrated in Figure 9. This effect provides a better disturbance rejection control than the classical PID control loop. Due to the amplitude dependence of the MIT rule, contributions of the ML-MR PID-MIT-CGM control to the disturbance rejection performance begin to vanish in the vicinity of the set-point 0. Consequently, the system can respond slightly better for step-up and step-down disturbances at the simulation times of 65 and 85 s compared to the optimal PID loop proposed by Bendjeghaba et al.

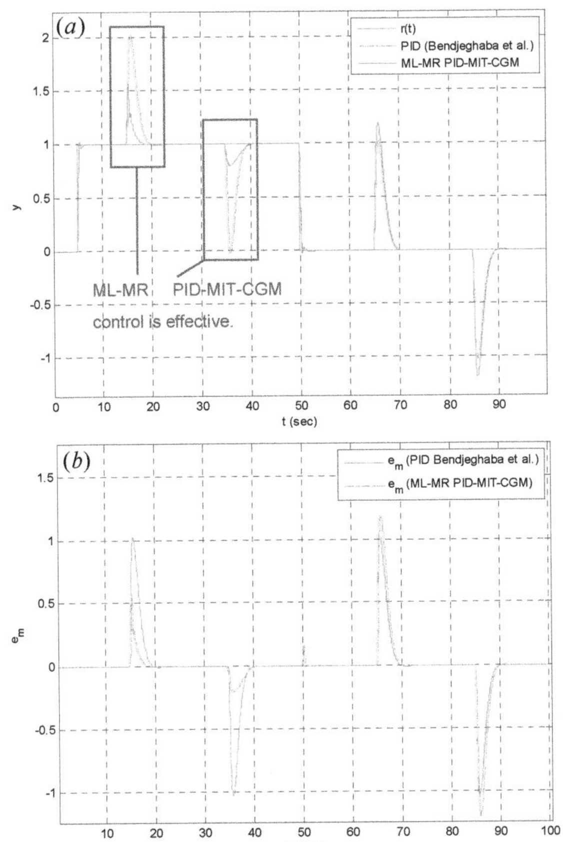

Figure 8. Improvement of the disturbance rejection performance of the classical optimal PID loop by the proposed ML-MR PID-MIT-CGM control structure for the AVR control system model. (**a**) Step response and (**b**) change of model error.

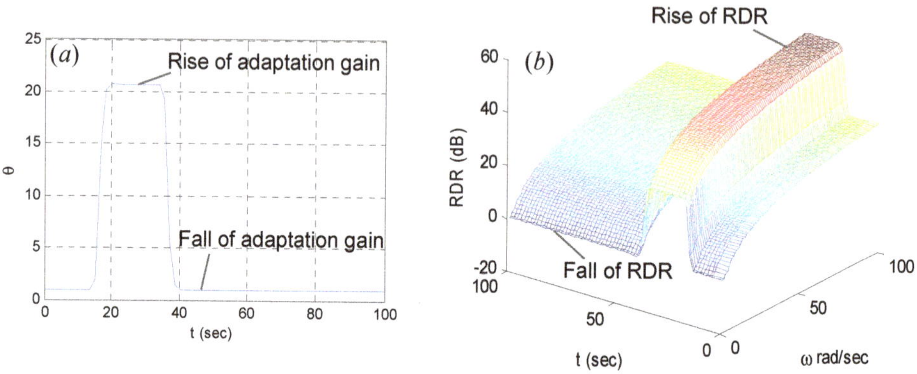

Figure 9. (**a**) Evolution of the adaptation gain θ of the ML-MR PID-MIT-CGM control structure. (**b**) Corresponding adaptive modification of the RDR spectrum of the $C_\theta(s)$ controller in order to reject disturbances.

Figure 9 describes how ML-MR PID-MIT-CGM control responds to disturbance incidents. For a disturbance incident, the gain θ increases, as shown in Figure 9b. This results in an increase of PID coefficients, and accordingly an increase in the RDR rates in the spectrum, as shown in Figure 9b. Therefore, the RDR performance of the optimal PID control loop of Bendjeghaba et al. is adaptively modified by adaptation gain θ, while the disturbances affect the system response. This figure verifies an important asset of the proposed ML-MR PID-MIT-CGM control structure: It adaptively adjusts the disturbance rejection control performance and retains the set-point control performance of the optimal PID loops in absence of disturbance incidents. Thus, it overcomes the inherent shortcoming that is caused by the performance tradeoff between set-point control and disturbance rejection control. This property can significantly contribute to the practical robust control performance of classical control loops and can be viewed as an elegant solution for increasing the disturbance rejection performance without deteriorating the set-point performance of control systems. However, there will be a few drawbacks in the PID-MIT structures for linear control system applications. The nonlinearity of the MIT rule leads to dependence of disturbance rejection performance on the level of the set-point. Another drawback for linear control applications is that the performance of the PID-MIT-CGM control structure is very sensitive to the time delay of plants. A sufficiently large time delay may easily lead to instability in the systems.

As mentioned in the previous section, the MIT rule does not behave linearly due to the terms of $y_m e_m$. To eliminate this nonlinearity and to make structural contributions to the PID control loop independent of the level of signal amplitudes, one can update the term $y_m e_m$ as ce_m, where the constant $c \in R$ can be typically set to 1. This modification implies that the sensitivity derivative is constant and can be shown as $\frac{de_m}{d\theta} = c$. Hence, Equation (16) is simplified to an integral operator and the adaptation gain can be written as:

$$\theta = \left|\gamma \frac{1}{s} e_m\right| \tag{18}$$

Using this modified adaptation gain—where Equation (18) replaces the MIT rule in Equation (16) —produces a new ML-MR control structure, which is called multi-loop model reference PID integral (ML-MR PID-I) control. The ML-MR PID-I control structure behaves linearly and improves the disturbance rejection performance independent of the reference input amplitude. Results in Figure 10 demonstrate this effect. Although ML-MR PID-MIT-CGM is very effective around the unity set-point, its performance degrades at set-point 0, although it is slightly better than the optimal PID loop proposed by Bendjeghaba et al. However, ML-MR PID-I control presents the same performance as ML-MR PID-MIT control at the unity set-point and can maintain this performance improvement at set-point 0. Figure 11 validates this effect by showing the adaptive adjustment of the RDR spectrum at set-point levels 1 and 0. As a consequence, for linear control systems with almost zero time delay, the proposed ML-MR PID-I control structure is capable of improving the disturbance rejection performance without deteriorating the set-point performance.

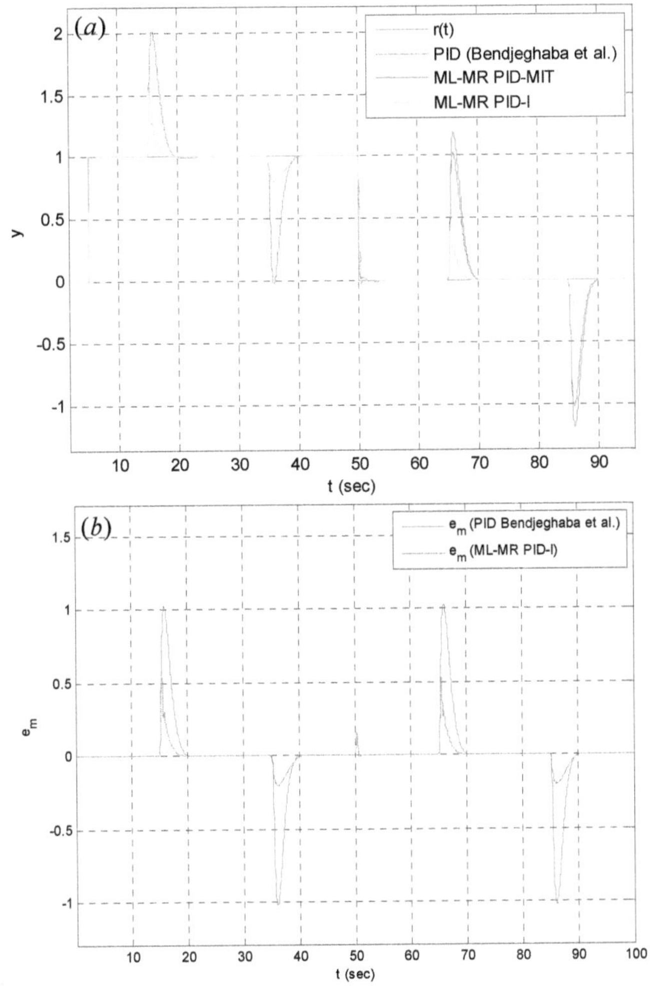

Figure 10. Improvement of the disturbance rejection performance of the classical optimal PID loop by the proposed ML-MR PID-I control structure for the model. (**a**) Step response and (**b**) change of model error.

The simulation results verify the effectiveness of the ML-MR control loops by showing that the ML-MR control has a performance capacity beyond the typical disturbance rejection of classical single-loop control loops. This is a very useful property that improves the robust control performance in the case of disturbance interference to the control system. The ML-MR PID-MIT-CGM control structure is more preferable for set-point control of nonlinear systems, where linear ML-MR control structures are not an effective solution.

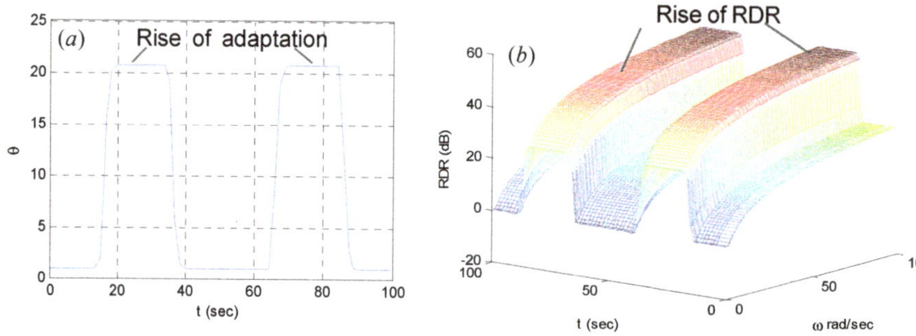

Figure 11. (**a**) Evolution of adaptation gain θ of the ML-MR PID-I control structure. (**b**) Adaptive modification of the RDR spectrum of the $C_\theta(s)$ controller for disturbance rejection.

4. Multi-Loop Model Reference PID Internal Model (ML-MR PID-IM) Control Structure: Internal Model Linear Adaptation Rule

Internal model control (IMC) relies on a mathematical model of a plant or process, which is referred to as an internal model and performs the control law according to this internal model. It was reported that IMC can be an effective solution for disturbance rejection and model perturbation problems [32]. In this section, we consider IMC to design a ML-MR control structure. Previously, an IMC-PID controller was proposed for improved disturbance rejection of time-delayed processes by using the reference model of the plant [14]. IMC law can be designed by manipulating transfer function models of the control system. If the internal model is an accurate enough representation of the system, the derived control law in the s-domain becomes valid for the real system.

In this section, we assume that the internal model is the PID control loop (the inner loop) and design an internal model PID controller, such that the outer loop can perform the IMC for the inner loop. Let us consider the multi-loop control structure that is depicted in Figure 1. To improve the disturbance rejection performance without influencing the control performance of the inner loop, the reference model, which is the internal model, is assumed to be the transfer function of the inner loop.

$$T_m(s) = T_1(s) = \frac{C_1(s)G(s)}{1 + C_1(s)G(s)} \tag{19}$$

One can write the transfer function that involves the internal model controller loop $C_2(s)$ and the theoretical reference model $T_1(s)$ as:

$$T_2(s) = \frac{C_2(s)T_1(s)}{1 + C_2(s)T_1(s)} \tag{20}$$

Our design objective is to find an internal model controller $C_2(s)$ so that the whole system behaves the same as the theoretical reference model $T_m(s)$, which represents an optimal design of the PID control loop. In order to achieve this objective, the transfer function condition $T_2(s) = T_m(s)$ should be satisfied. By using $T_m(s) = T_1(s)$ (Equation (19)), the transfer function condition becomes $T_2(s) = T_m(s) = T_1(s)$. It is used in Equation (20) to satisfy the transfer function condition, which gives:

$$T_1(s) = \frac{C_2(s)T_1(s)}{1 + C_2(s)T_1(s)} \tag{21}$$

Then, the controller function $C_2(s)$ is written as:

$$C_2(s) = \frac{1}{1 - T_1(s)} \tag{22}$$

By using Equation (19) in Equation (22), the $C_2(s)$ function is found:

$$C_2(s) = C_1(s)G(s) + 1 = L(s) + 1, \tag{23}$$

where the $L(s)$ function is the open loop transfer function of the inner loop, which is expressed as $L(s) = C_1(s)G(s)$. In this form, the $C_2(s)$ function can be directly implementable. Application steps for existing control loops are summarized as follows:

Step 1: The transfer function model of an optimal control loop is obtained, and this transfer function is used as the reference model ($T_m(s) = T_1(s)$).

Step 2: The optimal closed loop control system is enclosed by the outer loop, as shown in Figure 12.

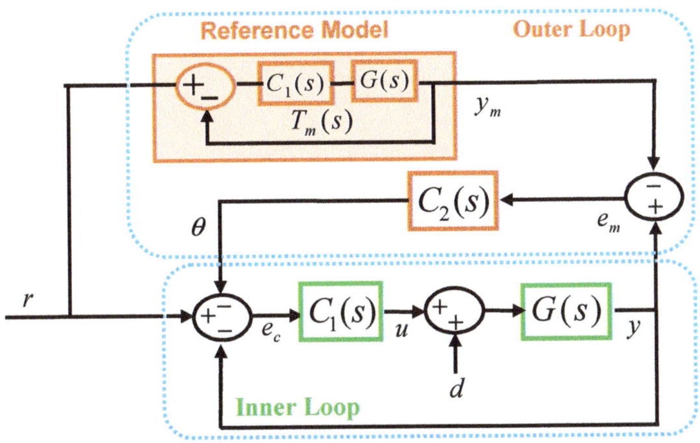

Figure 12. Block diagram of the ML-MR PID-IM control structure.

To demonstrate a possible application for the ML-MR PID-IM control, an example design for level control problem is presented. The plant model is an approximate model that represents the liquid level in the reboiler of the steam-heated distillation equipment [14,33]. The liquid level is controlled by adjusting the control valve on the steam line and the process model is given as $G(s) = \frac{-1.6(-0.5s+1)}{3s(s+1)}$ [14,33]. The optimal PID controller for improved disturbance rejection was given as [33]:

$$C_1(s) = -1.25 - \frac{0.2358}{s} - 1.8125s \tag{24}$$

The system model is unstable, and the controller function was obtained with negative coefficients. By considering Equation (23), the $C_2(s)$ function of the ML-MR PID-IM control structure is obtained as:

$$C_2(s) = \left(-1.25 - \frac{0.2358}{s} - 1.8125s\right)\frac{-1.6(-0.5s+1)}{3s(s+1)} + 1 \tag{25}$$

We implemented the ML-MR PID-IM control structure for this system, as shown in Figure 13, and performed control simulations for square waveform input disturbance.

Figure 14 shows responses of the ML-MR PID-IM control structure and Chen et al.'s optimal PID control loop. Results in Figure 14 clearly demonstrate that the ML-MR PID-IM control structure improves the disturbance rejection performance of Chen et al.'s optimal PID control loop without influencing its set-point performance. In the absence of disturbance, responses of the ML-MR PID-IM control structure overlap with responses of the optimal PID control loop. The set-point performance of the ML-MR PID-IM control is almost the same as the performance of the optimal PID control loop proposed by Chen et al. This is an important advantage for robust control performance so that the proposed ML-MR control structures can improve the disturbance rejection while maintaining the set-point performance of the existing PID control loops. A limitation of the ML-MR PID-IM control is that it is not an effective solution for large time delay plant dynamics and inaccurate plant models. High time delays can severely affect the performance of the ML-MR PID-IM control scheme.

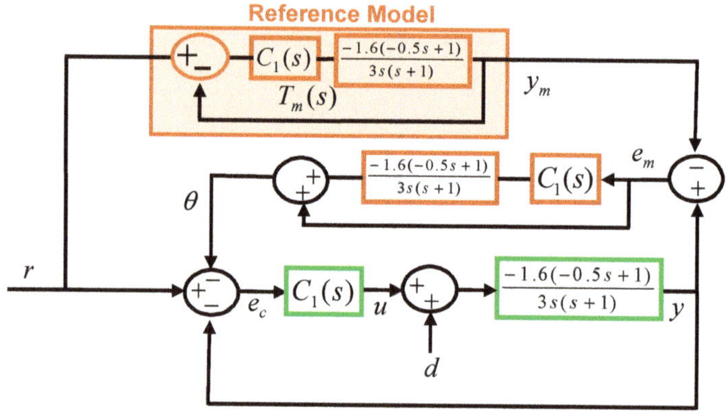

Figure 13. Implementation of the ML-MR PID-IM control structure for disturbance rejection control of the liquid level of the reboiler model.

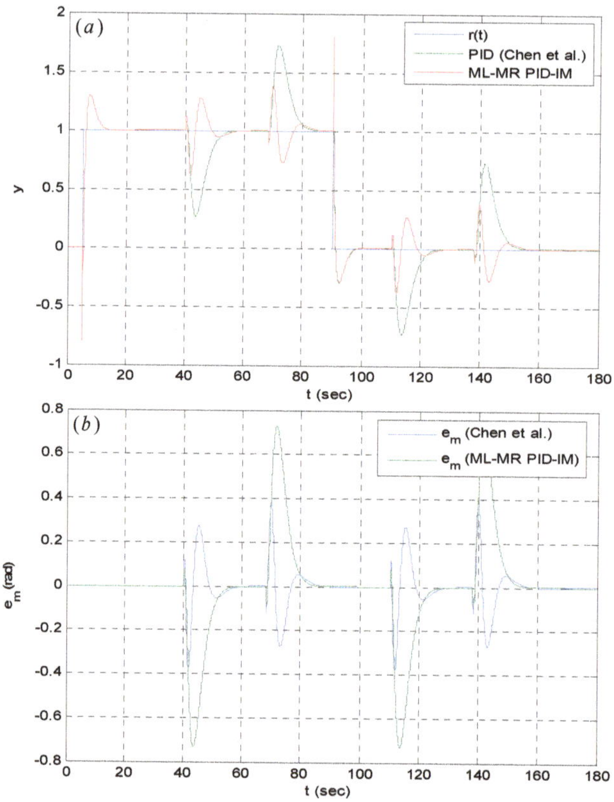

Figure 14. Improvement of the disturbance rejection performance of the classical optimal PID loop by the proposed ML-MR PID-IM control structure for liquid level control in the reboiler model. (**a**) Step response and (**b**) change of model error.

5. Multi-Loop Model Reference MIT-PID with Plant Function Adaptation (ML-MR MIT-PID-PFA): Nonlinear Adaptation Rule for Time-Delayed Systems

To improve the disturbance rejection capacity of time delay systems, model reference control is adopted for plant functions in a similar manner to classical IMC [14]. Any mismatch in responses from the real system and mathematical models of plants is assumed to be disturbance and is rejected by contributions of the MIT rule. In this configuration, the MIT rule forms the inner loop. The outer loop deals with the stability and set-point control. Therefore, this type of ML-MR control structure can be more effective for time delay systems because the adaptation loop only handles model mismatches between real plants and their mathematical models.

Given a first order plus dead time plant function in the form of $G(s) = \frac{K}{\tau s+1}e^{-Ls}$, an analytical design scheme for the ML-MR MIT-PID-PFA control structure can be suggested as follows.

Step 1: Identify a model of the plant function and take the reference model of the MRC loop as the plant function ($T_m(s) = G(s)$).

Step 2: Enclose the MRC loop with an optimal closed loop PID controller loop, as shown in Figure 15. Use the Tavakoli–Tavakoli PID tuning rule, which is expressed by Equation (10), for optimal operation of the PID controller function $C(s)$.

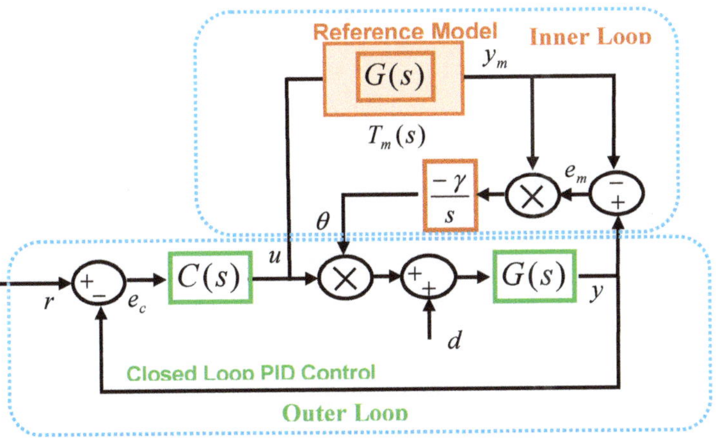

Figure 15. Block diagram of the ML-MR MIT-PID-PFA control structure.

An example design for the first order plus time delay plant model is given in the following section.

6. Multi-Loop Model Reference PID-PID with Plant Function Adaptation (ML-MR PID-PID-PFA): Linear Adaptation Rule for Time-Delayed Systems

To obtain a linear adaptation rule, the MIT rule, which leads to nonlinearity, is changed to classical PID controller. ML-MR PID-PID-PFA control is implemented, as illustrated in Figure 16.

The design steps of the ML-MR PID–PID-PFA control structure for a first order time delay plant in the form of $G(s) = \frac{K}{\tau s+1}e^{-Ls}$ can be carried out as follows.

Step 1: Identify the model of the plant function and take the reference model of the MRC loop as the plant function ($T_m(s) = G(s)$).

Step 2: Enclose the MRC loop with the optimal closed PID controller loop, as shown in Figure 16. Tavakoli-Tavakoli PID tuning rule, which is expressed by Equation (10), is used for the both PID controller functions.

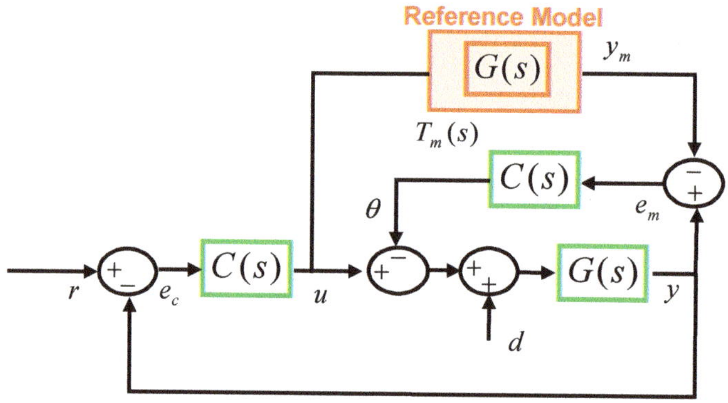

Figure 16. Block diagram of ML-MR PID-PID-PFA control structure.

To demonstrate the effectiveness of the ML-MR MIT-PID-PFA and ML-MR PID-PID-PFA control structures, we used them to enhance the first order plus deadtime plant model, which was given by Monje et al. [34].

$$G(s) = \frac{3.13}{433.33s + 1}e^{-50s} \tag{26}$$

This plant function is a first order dynamical model of the Basic Process Rig 38–100 Feedback Unit (produced by Feedback Instruments Ltd) experimental platform [34]. This system presents a 50 s time delay with a time constant of 433.33 s and is considered a large time delay process. Monje et al. designed an optimal fractional order PID (FOPID) controller for this system as follows:

$$C_{FOPID}(s) = 0.6152 + \frac{0.01}{s^{0.8968}} + 4.3867s^{0.4773} \tag{27}$$

We implemented the ML-MR MIT-PID-PFA and ML-MR PID-PID-PFA control structures for this system by following the design steps given above. For control performance comparison purposes, we implemented a classical optimal PID control system with the Tavakoli–Tavakoli PID tuning rule ($k_p = 1.1862$, $k_i = 0.0026$, $k_d = 22.8885$), a classical optimal PID control loop with MATLAB optimal tuning ($k_p = 0.8029$, $k_i = 0.002$, $k_d = -35.3930$), and an optimal FOPID control system with Monje et al.'s tuning rule. The set-point of this experimental system was originally 0.47 [34] and a step disturbance with an amplitude of 0.47 at 1000 s was applied to the input of the plant model. The results of the control system simulations are shown in Figure 17. The simulation results reveal that both the set-point and disturbance rejection performances of the ML-MR PID-PID-PFA structure re better than other controllers. The lowest overshoot and fastest settling results were achieved by the ML-MR PID-PID-PFA structure. It is noteworthy to observe that the control performance of the ML-MR PID-PID-PFA structure is superior to the optimal FOPID control system. This finding is a clear indication of the potential for multi-loop PID controllers to surpass the control performance of the FOPID controllers. The FOPID controllers have been considered as substitutes for classical PID control and have been extensively studied in the last decade, showing performance improvements of FOPID controllers over classical PID controllers. A major complication in practice for FOPID controllers is their realization complexity. Fractional order controllers can be implemented by using integer order models because near-ideal realization of fractional order derivatives and integral elements in digital systems is highly computationally expensive [2]. Recently, analog circuit realizations of fractional order controllers have been shown, with results promising low-cost and low-complexity solutions for practical realization of FOPID controllers [35,36]. Recent works have shown the robust control performance improvement of MR-ML control structures with FOPID controllers [25,26], indicating an avenue for robust control system research.

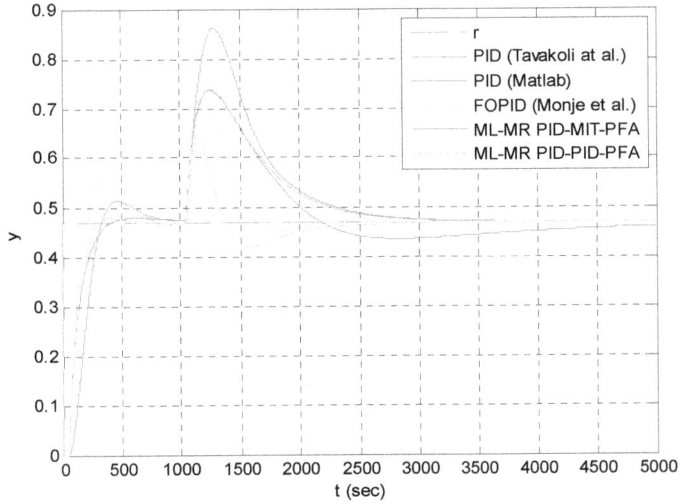

Figure 17. Control performance of various control systems for robust performance optimal control of a large time delay plant model [34].

It is useful to investigate the frequency dependence of disturbance rejection performance. For this purpose, the performances of the classical closed loop PID control structure and ML-MR PID-PID-PFA control structure were compared in the control problem of the Basic Process Rig 38–100 Feedback Unit experimental platform [34] using sinusoidal waveform disturbance signals at several frequencies. The PID controller was tuned according to the ITAE design rule proposed by Tavakoli et al. Table 1 shows the mean absolute error (MAE) of the control error signal ($\frac{1}{T}\int_{0}^{T}|e_c(t)|dt$) from simulations. For step disturbance, the ML-MR control structure reduces 50% the MAE of classical PID control system. In general, environmental disturbances are altered slowly and involve low frequency components. For very low frequency disturbance signals, the MAE of the ML-MR PID-PID-PFA control structure decreases down to 25% of the MAE of the classical PID control system. As the disturbance frequency increases, the MAE performance of the ML-MR PID-PID-PFA control structure equals the classical PID control structure. Figure 18 shows some of the simulation results and confirms these findings.

Table 1. Mean absolute error (MAE) of control error signals obtained from control simulations for step and sinusoidal waveform disturbances.

Angular Frequencies ω (rad/s)	Step Disturbance	0.001	0.005	0.01	0.05	0.1	0.5
Classical PID Control (Tavakoli et al.)	0.0470	0.0868	0.1748	0.1605	0.0560	0.0267	0.0145
ML-MR PID-PID-PFA	0.0234	0.0216	0.1037	0.1465	0.0690	0.0250	0.0140

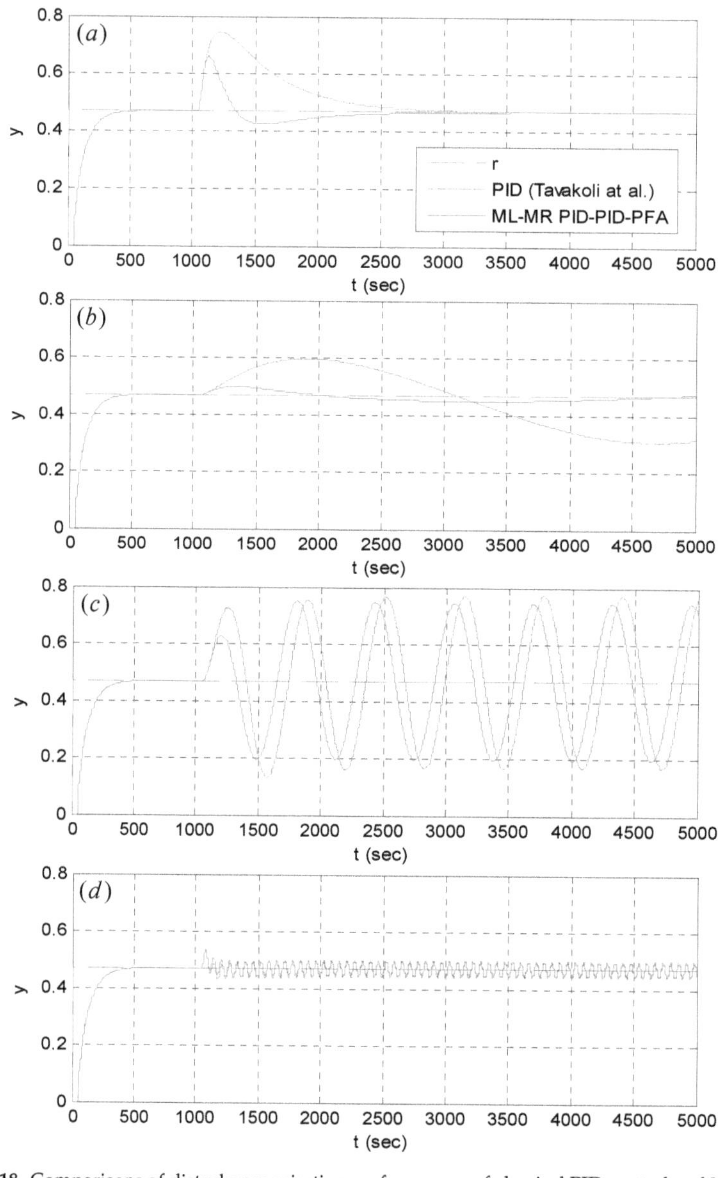

Figure 18. Comparisons of disturbance rejection performances of classical PID control and ML-MR PID-PID-PFA control structures for (**a**) step disturbance, (**b**) sinusoidal disturbance with a frequency of 0.001 rad/s, (**c**) sinusoidal disturbance with a frequency of 0.01 rad/s, and (**d**) sinusoidal disturbance with a frequency of 0.1 rad/s.

In former studies, a ML-MR control structure using the MIT rule was experimentally verified for control of an experimental magnetic levitation system [26] and control of an experimental electrical rotor [28]. We conducted an experimental study to validate the contribution of the ML-MR PID-PID-PFA control structure to the disturbance rejection performance of classical PID control systems. A twin-rotor multi-input multi-output (MIMO) system (TRMS) is a popular rotor control test platform that is preferred for electrical rotor control experiments [37,38]. The TRMS is composed of two electrical rotors [39]. These are pitch and yaw rotors, as illustrated in Figure 19. The angles of the rotors are controlled by regulating the input voltages of the DC electric motors. This adjusts the rotational speed of the propellers so that the rotors can hover.

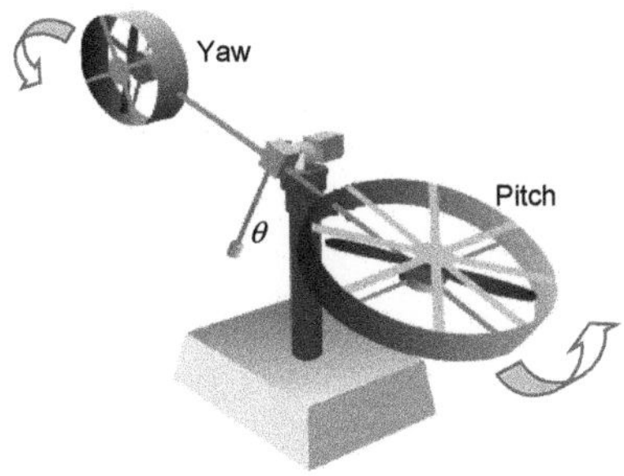

Figure 19. Experimental twin-rotor multi-input multi-output system (TRMS) setup [39].

Figure 20 shows the experimental results. After settling the pitch angle to 0.5 rad, an input disturbance with a unit step waveform was applied at 25 s. One can observe that the step-point control performances of the classical PID controller and ML-MR PID-PID-PFA control structure are almost the same. The slight differences are mainly caused by internal system noise and limitations of the reference model in representing whole dynamics in the main rotor of the TRMS experimental setup. When disturbance is applied to the system, the response of ML-MR PID-PID-PFA control structure differs from the response of the PID controller and better rejects the disturbance at the system output, as shown in the figure.

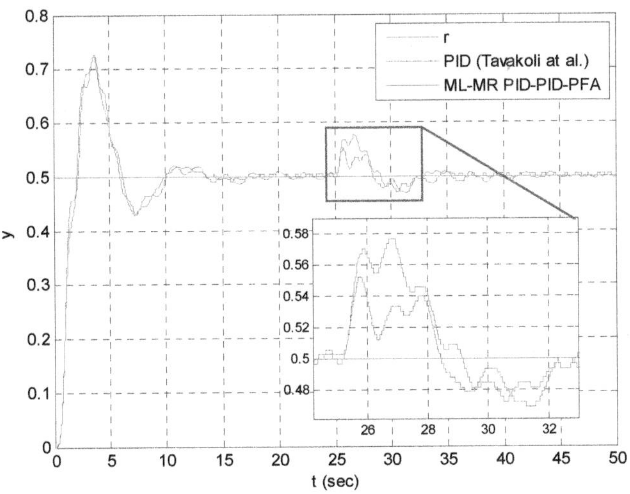

Figure 20. Experimental results obtained for control of the main rotor of TRMS setup.

One of major technical complications that was observed in the experimental study was the model mismatch problem in the real system. We used a nonlinear model of TRMS that is a good representation of experimental systems. However, inherent limitations of mathematical modeling and changes of operating conditions reduce the model consistency and lead to model mismatching problems. Since the response of reference model does not adequately match with the response of the real TRMS setup, the PID controller of the adaptation loop, which generates the adaptation gain θ, can cause instability in the system for the same PID controller coefficients for the control loop. This stability problem can be solved by retuning the PID controller to the adaptation loop.

7. Discussions and Conclusions

This study investigated some contemporary ML-MR adaptive control structures that are applicable for classical PID control loops. These structures are the ML-MR PID-MIT, ML-MR PID-MIT-CGM, ML-MR PID-I, ML-MR PID-IM, ML-MR MIT-PID-PFA, and ML-MR PID-PID-PFA control structures. Design problems of these ML-MR structures were addressed, and straightforward design and implementation strategies were presented. The performance evaluations, advantages, and drawbacks of these ML-MR structures were discussed according to control simulation results. Simulation results indicated significant disturbance rejection performance improvements in four different control application examples. The performance comparisons with several single-loop optimal control loops theoretically show that improvement of the disturbance rejection control without affecting the set-point performance is possible with presented ML-MR control structures. Some noteworthy properties of the presented ML-MR adaptive control structures are listed in Table 2.

Table 2. Some properties of the presented ML-MR control structures.

ML-MR Control Structures	Character	Dependence to Set-Point Level	Negative Effects on Set-Point Performance	Applicability for Time Delay Systems	Indications in Simulation Results
ML-MR PID-MIT	Nonlinear	Dependent	None	None	Figures 3 and 4
ML-MR PID-MIT-CGM	Nonlinear	Dependent	None	None	Figure 8
ML-MR PID-I	Linear	Independent	None	None	Figure 10
ML-MR PID-IM	Linear	Independent	None	None	Figure 14
ML-MR MIT-PID-PFA	Nonlinear	Dependent	None	Applicable to some degree	Figure 17
ML-MR PID-PID-PFA	Linear	Independent	None	Applicable to some degree	Figure 17

Some notable remarks: * ML-MR adaptive control structures are applicable in all classical control loops in order to improve the disturbance rejection performance without modifying the parameters of the existing control system or degrading the set-point performance. This is an important advantage of hierarchical ML-MR structures in control engineering. * Integration approaches of the model reference control loop into classical control loops are carried out in two steps: outer loop integration uses closed loop control systems as reference models, such as ML-MR PID-MIT, ML-MR PID-MIT-CGM, ML-MR PID-I, and ML-MR PID-IM controls; and inner loop integration uses models of plant or process functions as reference models, such as ML-MR MIT-PID-PFA and ML-MR PID–PID-PFA. Outer loop integrations cover the whole closed control loop to deal with disturbances, and several works have indicated that these type integrations are preferable solutions for improving fault tolerance and disturbance rejection in the whole control loop [25–28]. Inner loop integration only encloses the plant function, and hence it can be employed to deal with external disturbances or parametric perturbations that influence the controlled system. * ML-MR PID-MIT-CGM and ML-MR PID-I structures can adaptively increase the RDR performance of the closed control loops without affecting set-point performance. The ML-MR PID-IM control structure benefits from having an internal model controller design, however it should be noted that its performance is very sensitive to the accuracy of plant and controller models. * Large-time delay systems that set the point and disturbance rejection performance of ML-MR PID-PID-PFA control structures can surpass the disturbance rejection performance of a single-loop optimal FOPID control system. These results reveal that the disturbance rejection capacity of closed loop PID control loops can be exceeded by combining them with ML-MR control structures. ML-MR control structures are a feasible way to resolve performance tradeoff between set-point and disturbance rejection control structures. * Stability analysis of linear ML-MR control structures is relatively straightforward and is performed by obtaining the transfer function of the whole structure and checking for pole placements for left half-plane (LHP) stability. The satiability of a nonlinear ML-MR structure, such as ML-MR PID-MIT control, along with the stability and convergence conditions, are derived on the basis of the bounded input and bounded output stability state of the model error (see Appendix A). * Results in this study are based on simulation models. The reference models perfectly match the controlled systems. It is very important to note that model mismatches due to limitations in mathematical modeling and parametric model perturbations in real systems may have severe effects on the control performance and stability of the presented ML-MR adaptive control structures in real-world applications. The sensitivity of these structures to model mismatching should be carefully investigated in future works. The fragility of the system stability due to model imprecision should be carefully considered when designing real control systems. Moreover, the resolution of sensors and I/O ports of control cards are important in practical applications. * This study is a theoretical demonstration of six different ML-MR adaptive control structures. Future works should address experimental performance evaluations and practical realization issues of ML-MR adaptive control structures.

The most prominent contributions of the introduced hierarchical ML-MR control structures are:

(i) The proposed ML-MR adaptive control structures can be applied to existing control loops without modifying any parameters of the control loops. These structures do not require online return of the control loop, which may result in instant performance degradation while altering controller coefficients. Therefore, ML-MR control structures provide more consistent control performance than conventional MRAC structures that are used to perform online return of the control loop.

(ii) The proposed ML-MR adaptive control structures increase the disturbance rejection performance without deteriorating the set-point control quality. This is an important contribution to the solution of performance tradeoffs between disturbance rejection and set-point control.

Author Contributions: Conceptualization, B.B.A. and A.T.; methodology, B.B.A. and A.T.; validation, E.P. and C.Y.; writing—original draft preparation, B.B.A.; writing—review and editing, A.T., E.P., and C.Y.; supervision, E.P. and C.Y. All authors have read and agreed to the published version of the manuscript.

Funding: This research was funded by the European Cooperation in Science and Technology (COST) Action CA15225. The work of Eduard Petlenkov and Aleksei Tepljakov was supported by the Estonian Research Council grant PRG658.

Acknowledgments: This article is based upon work from COST Action CA15225, a network supported by the European Cooperation in Science and Technology (COST). The work of Eduard Petlenkov and Aleksei Tepljakov was supported by the Estonian Research Council grant PRG658.

Conflicts of Interest: The authors declare no conflict of interest. The funders had no role in the design of the study; in the collection, analyses, or interpretation of data; in the writing of the manuscript, or in the decision to publish the results.

Appendix A

Since the inner PID control loop is designed as a stable system, convergence of the model error $e_m = y - y_m < \varepsilon \in R$ ensures the bounded input and bounded output stability of the ML-MR PID-MIT control structure. In this manner, we can investigate the convergence condition of e_m.

Lemma A1. *For the ML-MR PID-MIT control structure, the model error becomes zero ($e_m = 0$) when the adaptation gain θ is equal to 1 and $T(s) = T_m(s)$.*

Proof A1. Let us consider model error $e_m = y - y_m = T(s)\theta r - T_m(s)r$ for the ML-MR PID-MIT control structure and apply the condition of $\theta = 1$ and $T(s) = T_m(s)$. Accordingly, one can show that:

$$e_m = T(s)\theta r - T_m(s)r = T_m(s)r - T_m(s)r = 0$$

□

Theorem A1 (Stability of the ML-MR PID-MIT Control Structure). *For a stable system model $T(s) = T_m(s)$, the model of the ML-MR PID-MIT control structure is stable only if the model error function is as follow:*

$$e_m = -\frac{1}{\gamma}\left(s^2 + \frac{s^2}{C(s)G(s)}\right), \tag{A1}$$

where the bounded input and bounded output are stable. For the below equation the model error converges to zero.

$$\lim_{s \to 0^+}\left(s^3 + \frac{s^3}{C(s)G(s)}\right) = 0, \tag{A2}$$

Proof A2. Let us consider the adaptation rule of the ML-MR PID-MIT control structure, which is given by Equation (9) as $\theta = -\gamma\frac{1}{s}y_m e_m$. The model can be written as:

$$e_m = \frac{s\theta}{-\gamma y_m} \quad \text{(A3)}$$

The output of the reference model y_m can be written as $y_m = T_m(s)r(s)$. For set-point control stability, one considers the step response of the system. Therefore, the reference input is assumed to be a step function by using $r(s) = \frac{1}{s}$ in Equation (A3).

$$e_m = \frac{s^2\theta}{-\gamma T_m(s)} \quad \text{(A4)}$$

According to Lemma 1, the convergence of the model error to zero ($e_m = 0$) is possible for $\theta = 1$ in the case of $T_m(s) = T(s)$. By applying this condition, the model error is written as:

$$e_m = \frac{s^2}{-\gamma T(s)} \quad \text{(A5)}$$

The function $T(s)$ is the transfer function of the inner loop, which is stable due to optimal tuning of the PID control system. When the $T(s)$ function given by Equation (2) is used in Equation (A5), one obtains

$$e_m = -\frac{1}{\gamma}\left(s^2 + \frac{s^2}{C(s)G(s)}\right) \quad \text{(A6)}$$

The inner loop is configured to be stable, therefore the output of the inner loop is bounded for $y < \xi \in R$ for a step reference input. When the model error e_m of the system given by Equation (A6) is shown to be stable, the bounded input and bounded output of the ML-MR PID-MIT control structure become stable, so that $e_m = y - y_m < \varepsilon \in R$ is satisfied. In order to ensure convergence of the model error to zero, a final value theorem can be applied as:

$$\lim_{s\to 0^+} s e_m = 0 \quad \text{(A7)}$$

Then, one obtains the convergence condition as follows:

$$\lim_{s\to 0^+}\left(s^3 + \frac{s^3}{C(s)G(s)}\right) = 0. \quad \text{(A8)}$$

□

References

1. Fossard, A.J.; Normand-Cyrot, D. *Nonlinear Systems Modeling and Estimation*; Springer: Berlin, Germany, 1995.
2. Tepljakov, A. *Fractional-Order Modeling and Control of Dynamic Systems*; Springer: Berlin, Germany, 2017.
3. Lanchier, N. *Stochastic Modeling*; Springer: Berlin, Germany, 2017.
4. Alagoz, B.B.; Tan, N.; Deniz, F.N.; Keles, C. Implicit disturbance rejection performance analysis of closed loop control systems according to communication channel limitations. *IET Control Theory A* **2015**, *9*, 2522–2531. [CrossRef]
5. Butler, H.; Honderd, G.; Van Amerongen, J. Model reference adaptive control of a direct-drive DC motor. *IEEE Control Syst. Mag.* **1989**, *9*, 80–84. [CrossRef]
6. Barkana, I.; Kaufman, H.; Balas, M. Model reference adaptive control of large structural systems. *J. Guid. Control Dyn.* **1983**, *6*, 112–118. [CrossRef]
7. Duarte, M.A.; Narendra, K.S. A new approach to model reference adaptive control. *Int. J. Adapt. Control.* **1989**, *3*, 53–73. [CrossRef]

8. Barkana, I. Simple adaptive control–a stable direct model reference adaptive control methodology–brief survey. *Int. J. Adapt. Control* **2014**, *28*, 567–603. [CrossRef]
9. Yamamoto, T.; Takao, K.; Yamada, T. Design of a data-driven PID controller. *IEEE Trans. Control Syst. Technol.* **2008**, *17*, 29–39. [CrossRef]
10. Barbosa, R.S.; Machado, J.T.; Ferreira, I.M. Tuning of PID controllers based on Bode's ideal transfer function. *Nonlinear Dyn.* **2004**, *38*, 305–321. [CrossRef]
11. Wellenreuther, A.; Gambier, A. Multi-loop Controller Design for a Heat Exchanger. In Proceedings of the 2006 IEEE International Conference on Control Applications, Munich, Germany, 4–8 October 2006; pp. 2099–2104.
12. Vijayan, V.; Panda, R.C. Design of PID controllers in double feedback loops for SISO systems with set-point filters. *ISA Trans.* **2012**, *51*, 514–521. [CrossRef]
13. Lee, M.; Lee, K.; Kim, C.; Lee, J. Analytical Design of Multiloop PID Controllers for Desired Closed-Loop Responses. *AIChE J.* **2004**, *50*, 1631–1635. [CrossRef]
14. Shamsuzzoha, M.; Lee, M. IMC-PID Controller Design for Improved Disturbance Rejection of Time-Delayed Processes. *Ind. Eng. Chem. Res.* **2007**, *46*, 2077–2091. [CrossRef]
15. Visioli, A. A new design for a PID plus feedforward controller. *J. Process Control* **2004**, *14*, 457–463. [CrossRef]
16. Su, W.A. Model Reference-Based Adaptive PID Controller for Robot Motion Control of Not Explicitly Known Systems. *IJICS* **2007**, *12*, 237–244.
17. Trajkov, T.N.; Köppe, H.; Gabbert, U. Direct model reference adaptive control (MRAC) design and simulation for the vibration suppression of piezoelectric smart structures. *Commun. Nonlinear Sci.* **2008**, *13*, 1896–1909. [CrossRef]
18. Landau, I.D. *Adaptive Control the Model Reference Approach*; Marcel Dekker: New York, NY, USA, 1979.
19. Astrom, K.J.; Wittenmark, B. *Adaptive Control*; Addison-Wesley: Boston, MA, USA, 1995.
20. Butler, H. *Model-Reference Adaptive Control-From Theory to Practice*; Prentice-Hall: Upper Saddle River, NJ, USA, 1992.
21. Vinagre, B.M.; Petrás, I.; Podlubny, I.; Chen, Y.Q. Using fractional order adjustment rules and fractional order reference models in model-reference adaptive control. *Nonlinear Dyn.* **2001**, *29*, 269–279. [CrossRef]
22. Kavuran, G.; Alagoz, B.B.; Ates, A.; Yeroglu, C. Implementation of model reference adaptive controller with fractional order adjustment rules for coaxial rotor control test system. *BAJECE* **2016**, *4*, 84–88.
23. Kavuran, G.; Ates, A.; Alagoz, B.B.; Yeroglu, C. An Experimental Study on Model Reference Adaptive Control of TRMS by Error-Modified Fractional Order MIT Rule. *Control Eng. Appl. Inf.* **2017**, *19*, 101–111.
24. Benaskeur, A.R.; Desbiens, A. Backstepping-based adaptive PID control. *IET Control Theory A* **2002**, *149*, 54–59.
25. Alagoz, B.B.; Tepljakov, A.; Petlenkov, E.; Yeroglu, C. Multi-Loop model reference adaptive control of fractional-order PID control systems. In Proceedings of the 2017 40th International Conference on Telecommunications and Signal Processing (TSP), Barcelona, Spain, 5–7 July 2017; pp. 702–705.
26. Tepljakov, A.; Alagoz, B.B.; Gonzalez, E.; Petlenkov, E.; Yeroglu, C. Model reference adaptive control scheme for retuning method-based fractional-order PID control with disturbance rejection applied to closed-loop control of a magnetic levitation system. *J. Circuits Syst. Comput.* **2018**, *27*, 1850176. [CrossRef]
27. Rajesh, R.; Deepa, S.N. Design of direct MRAC augmented with 2 DoF PIDD controller: An application to speed control of a servo plant. *J. King Saud Univ. Eng. Sci.* **2019**. [CrossRef]
28. Alagoz, B.B.; Kavuran, G.; Ateş, A.; Yeroğlu, C.; Alisoy, H. Multi-loop Model Reference Adaptive PID Control for Fault-Tolerance. *BAJECE* **2019**, *7*, 276–285. [CrossRef]
29. Tavakoli, S.; Tavakoli, M. Optimal tuning of PID controllers for first order plus time delay models using dimensional analysis. In Proceedings of the 2003 4th International Conference on Control and Automation Proceedings, Montreal, QC, Canada, 12 June 2003; pp. 942–946.
30. Alagoz, B.B.; Deniz, F.N.; Keles, C.; Tan, N. Disturbance rejection performance analyses of closed loop control systems by reference to disturbance ratio. *ISA Trans.* **2015**, *55*, 63–71. [CrossRef] [PubMed]
31. Bendjeghaba, O.; Boushaki, S.I. Optimal Tuning of PID Controller in Automatic Voltage Regulator System using Improved Harmony Search Algorithm. In Proceedings of the 7th Global Conference on Power Control and Optimization, Prague, Czech, 25–27 August 2013; pp. 1–6.
32. Zhang, P. Industrial control system simulation routines. In *Advanced Industrial Control Technology*; Elsevier: Oxford, UK, 2010; pp. 781–810.

33. Chen, D.; Seborg, D.E. PI/PID Controller Design Based on Direct Synthesis and Disturbance Rejection. *Ind. Eng. Chem. Res.* **2002**, *41*, 4807–4822. [CrossRef]
34. Monje, C.A.; Vinagre, B.M.; Feliu, V.; Chen, Y.Q. Tuning and auto-tuning of fractional order controllers for industry applications. *Control Eng. Pract.* **2008**, *16*, 798–812. [CrossRef]
35. Muñiz-Montero, C.; García-Jiménez, L.V.; Sánchez-Gaspariano, L.A.; Sánchez-López, C.; González-Díaz, V.R.; Tlelo-Cuautle, E. New alternatives for analog implementation of fractional-order integrators, differentiators and PID controllers based on integer-order integrators. *Nonlinear Dyn.* **2017**, *90*, 241–256. [CrossRef]
36. Domansky, O.; Sotner, R.; Langhammer, L.; Jerabek, J.; Psychalinos, C.; Tsirimokou, G. Practical design of RC approximants of constant phase elements and their implementation in fractional-order PID regulators using CMOS voltage differencing current conveyors. *Circuit Syst. Signal Process.* **2019**, *38*, 1520–1546. [CrossRef]
37. Alagoz, B.B.; Ates, A.; Yeroglu, C. Auto-tuning of PID controller according to fractional-order reference model approximation for DC rotor control. *Mechatronics* **2013**, *23*, 789–797. [CrossRef]
38. Rahideh, A.; Shaheed, M.H. Constrained output feedback model predictive control for nonlinear systems. *Control Eng. Pract.* **2012**, *20*, 431–443. [CrossRef]
39. *Twin Rotor MIMO System Control Experiments 33-949S*; Feedback Instruments Ltd.: East Sussex, UK.

algorithms

MDPI

Article

Optimal Learning and Self-Awareness Versus PDI

Brendon Smeresky [1], Alex Rizzo [1] and Timothy Sands [2,*]

1 Department of Mechanical and Aerospace Engineering, Naval Postgraduate School, Monterey, CA 93943, USA; bpsmeres@nps.edu (B.S.); akrizzo@nps.edu (A.R.)
2 Department of Mechanical Engineering (CVN), Columbia University, New York, NY 10027, USA
* Correspondence: dr.timsands@caa.columbia.edu; Tel.: +1-831-656-3954

Received: 26 November 2019; Accepted: 9 January 2020; Published: 11 January 2020

Abstract: This manuscript will explore and analyze the effects of different paradigms for the control of rigid body motion mechanics. The experimental setup will include deterministic artificial intelligence composed of optimal self-awareness statements together with a novel, optimal learning algorithm, and these will be re-parameterized as ideal nonlinear feedforward and feedback evaluated within a Simulink simulation. Comparison is made to a custom proportional, derivative, integral controller (modified versions of classical proportional-integral-derivative control) implemented as a feedback control with a specific term to account for the nonlinear coupled motion. Consistent proportional, derivative, and integral gains were used throughout the duration of the experiments. The simulation results will show that akin feedforward control, deterministic self-awareness statements lack an error correction mechanism, relying on learning (which stands in place of feedback control), and the proposed combination of optimal self-awareness statements and a newly demonstrated analytically optimal learning yielded the highest accuracy with the lowest execution time. This highlights the potential effectiveness of a learning control system.

Keywords: control systems; feedforward; feedback; learning systems; deterministic artificial intelligence; Luenberger; proportional-derivative-integral; PDI; virtual zero reference; dead-beat control inspired

---

1. Introduction

The goal of rotational mechanics control is to have a system that can move to and hold a specific orientation in three-dimensional space, relative to an inertial frame. The term system generically applies to many physical practices, and aerospace systems are emphasized here with extension to maritime systems by description. The goal may be viewed through three different lenses: classical control, modern control, and/or artificial intelligence (either stochastic or deterministic). These lenses explain the same control theory in three different contexts. For all three paradigms, consideration of motion mechanics must include kinetics, kinematics, disturbances, controls, actuators, and estimators that dictate the system's motion [1]. Specifically, with regard to classical control, both feed-forward and feedback controller are implemented in order to eliminate error between a desired and commanded signal [2]. With regard to modern control, the classical notion of feedforward and feedback is contemplated in terms of an estimation [3] and correction method [4–6] implemented using a non-linear control estimator coupled with a nonlinear corrector in order to reduce error. The third context relates control systems to deterministic artificial intelligence and machine learning.

In today's world, machine learning and artificial intelligence are usually referred as same. But that is not the truth, both are different in so many aspects. Machine learning is one of these aspects. Artificial intelligence is by far divided into two approaches, statistical and deterministic to program a machine to mimic human beings. When a machine is given the data to rely on for its intelligence, it's called the statistic or probabilistic approach. When a result is derived by the machine through a series of conditions, it's called deterministic approach [7].

1.1. The Contributions

The contributions to the field described in this manuscript include augmentation of feedback proportional-derivative-integral, PDI control (a modified form of proportional-integral-derivative, PID control) with a nonlinear de-coupling control that seeks to account for the nonlinear coupling of vector cross-products. The nominal PDI controller is tuned with the accepted methodologies [2,4–6], and the slightly novel approach is the augmentation. The augmented PDI control is compared to state-optimal feedforward control. A larger contribution is the development of deterministic self-awareness statements in lieu of feedforward control, while another substantial contribution is the development of optimal learning statements resulting from exact reparameterization of the nonlinear forms into a linear diffeomorphism. The proposed systems are all shown to very effectively drive rigid-body systems to desired configurations, while the combination of optimal self-awareness statements and optimal learning comprising deterministic artificial intelligence will be shown to be the most superior approach. Readers wishing to preview the claims before continuing to read the manuscript should refer Figure 8 to see the proposed approaches compared using very small-scale state errors.

1.2. The Literature Review

Ashford proposed a deterministic artificial intelligence approach for cyber information security relying on self-awareness for defense and healing [8]. Significant recent research has investigated the self-awareness of several animals [9–11], including elephants [12], magpies [13,14], dolphins [15], chimpanzees [16], ants [17], and humans [18,19], including children [20–24].

Scholars have long pondered the nature of self-awareness [25–27] and its neural basis in stochastic, non-deterministic approaches [28]. The emerging research in self-awareness of contexts includes using cross-domain computing environments to automatically identify the context of the user [29]. The environmental impacts on driving styles are predicted using new methods and models [30]. Another recent self-awareness work accounts for predictive capabilities, immersive devices, multimodal interactions, and adaptive displays of multi-system robots to develop interfaces including predictive virtual reality, conventional, and predictive conventional [31]. Assisting people to copy with incurable diseases including chronic obstructive pulmonary disease using context-aware systems has been extensively examined in publications of recent decades [32]. Context-awareness and social computing integrating multiple technologies has spawned a conceptual framework for collaborative context-aware learning activities [33].

Stochastically using both business and information services to monitor hazards while driving helps ensure driving safety [34]. Using large amounts of data sources, including sensors streaming, in manufacturing self-optimizing algorithms approaches context sensitivity [35]. Emotional components may augment context-awareness with the advent of simulations of empathy and emotion [36]. Information on the state and progress of systems of computing systems which can maintain models and learn leads to the ability for the systems to associate their behaviors with reason instantiating computational self-awareness [37]. This key novel work, only three years old, inspires the adoption of self-awareness as a new way of thinking (as a starting point for artificial intelligence), and it has spawned a recent lineage of the growth of the notion in typical stochastic (non-deterministic) approaches to artificial intelligences.

Wireless sensor networks were augmented by Kao, et al. with self-awareness paradigms providing autonomy and fault-tolerant adaptive routing overcoming limits of self-selective and self-healing routing [38]. Adaptive health monitoring was achieved by self-awareness for systems using information from five levels from the configuration level to the level of data connection. [39] Higher personality state variability was addressed by Jauk, et al. [40] using self-awareness seeking to explain contradictory results using the so-called "theory of mind" offered by Wundrack, et al. [41] Jauk's argument focused on Wundrack's isolation of the causality direction, and furthermore point out the ability of perspective-taking may be both cause and effect of personality state variability.

Heterogeneous or homogeneous types of potency of multi-robot systems were achieved by Kosak, et al. by integration of algorithms and mechanisms into both simulation and laboratory experiments [42]. Human behavior mimicking by robots with self-awareness and awareness of their environment proved able to autonomously operate in dynamic environments with optimal route planning [43].

Various instantiations of stochastic artificial intelligence (A.I.) described above seek optimality after acceptance of the structure of the system and form of learning. None of these are used in the deterministic artificial intelligence algorithm presented here, instead they serve as a background to highlight the distinction between stochastic and non-stochastic (deterministic) artificial intelligence [2–8]. Deterministic artificial intelligence instantiated here, stems from a lineage of nonlinear adaptive control [44–46] inspired learning augmented by physics-based control methods [47,48] inspired deterministic self-awareness statements. A rigid body's self-awareness of its own attributes is updated every time-step with new information from learning, noting the generic term "rigid-body" applies to physical systems in many disparate disciplines. The aforementioned lineage of references was applied to aerospace systems (both autonomous jet aircraft and spacecraft), while the novel methods presented here are simultaneously being applied to maritime systems to be revealed in sequel research being prepared for publication. Notice the algorithm presented here disposes of feedback control, replacing it with deterministic self-awareness statements that are parameterized such that learning algorithms use control-error feedback (not state feedback) to learn the proper, time-variant self-awareness statement. Notice the typical control calculation in Figure 1 composed of any number of kinds of feedback control are eliminated in Figure 2 will be replaced. The self-awareness statements necessitate online autonomous trajectory generation, done here in accordance with reference [49]. This self-updating learning mechanism can be viewed akin to the update cycle used by supervised learning algorithms to model a system's performance. As an example, the updating mechanism is either a linear or non-linear method to update an unknown inertia matrix for a rigid body [8–45], while new learning methods will be proven here to be optimal.

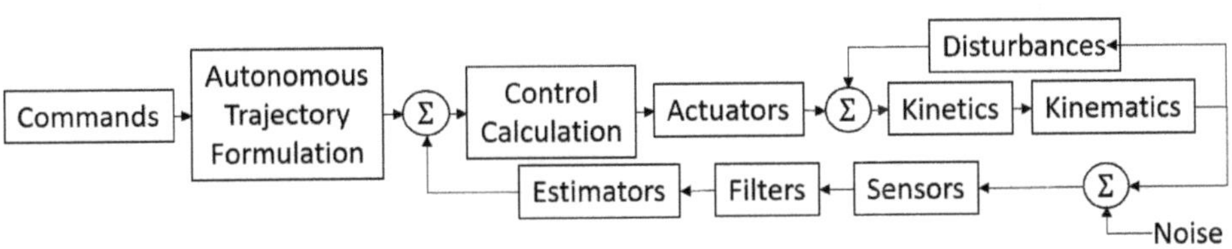

Figure 1. The system topology from commanded end-state and time, through autonomously calculated desired roll ϕ_d, pitch θ_d, and yaw ψ_d inputs, through control calculations executed by control moment gyroscope actuators resulting in actual Euler angle outputs of kinematic expressions of kinetic responses. Notice that actual responses are unknown, but sensed and filtered, and then used in state estimators to provide full state feedback.

Figure 1 depicts the topology of the computational steps that take desired angle inputs and calculates Euler Angle outputs: roll ϕ, pitch θ, and yaw ψ. The desired angle inputs are processed through the trajectory, controls, actuators, dynamics, and disturbances blocks. Section 2 will explain the theory behind the overall control system. Section 3 will detail the experimental setup and the results, and Section 4 will conclude this manuscript hopefully achieving the objective of this paper: introduce a new paradigm for self-awareness and optimal learning in the category deterministic artificial intelligence.

2. Materials and Methods

The rotation maneuver of a rigid body (representing at least aerospace and maritime systems) from one position to another is measured from the inertial reference frame or $\mathcal{F}_I \equiv \begin{bmatrix} \mathbb{X}_i & \mathbb{Y}_i & \mathbb{Z}_i \end{bmatrix}$ to the final position measured in the body reference frame annotated $\mathcal{F}_B \equiv \begin{bmatrix} \mathbb{X}_B & \mathbb{Y}_B & \mathbb{Z}_B \end{bmatrix}$. For the simulations presented here, a model was created to rotate in a user-prescribed maneuver time from an initial arbitrary orientation $\{\theta_i\} = \{\ \mathbb{X}_A \quad \mathbb{Y}_A \quad \mathbb{Z}_A\ \}^T$ to a final arbitrary orientation $\{\theta_f\} = \{\ \mathbb{X}_B \quad \mathbb{Y}_B \quad \mathbb{Z}_B\ \}^T$, where the orientations may also be set by the users. The kinetics and kinematics, accounting for motion in the orbital frame, and realistic disturbance calculations are all explained in reference [1]. This section of the manuscript focuses on deterministic artificial intelligence used for the control calculation utilizing an error estimation and correction mechanism. Simulations will be provided in Section 3 utilizing three control moment gyroscopes [50] that are responsible for physically moving the system according to the inputted control signal.

2.1. Rigid Body Mechanics

Rigid bodies rotate in accordance with Euler's moment equation, and this physics-based governing differential equation will be used later to formulate a control in a feedforward topology that asserts behavior in accordance with the fact the item being controlled is a rigid body. It is desired for the autonomous system to be self-aware that it is (for example) an aerospace or maritime system which will obey Euler's moment equation. This general notion will utilize the nomenclature self-awareness statement, since the control enforces the self-awareness that the control-item is a rigid body and must obey Euler's equations, and by this method eliminates the need for learning of the structure of the output data (instantiating a significant improvement over stochastic artificial intelligence methods). Equation (1) is Euler's moment equation for rigid body mechanics in three dimensions, one equation for each dimension in matrix form.

$$T = \dot{H} + \omega \times H_s = u^* \tag{1}$$

These governing equations apply to any rotating, rigid body of mass and thus it's applicability to maritime and aerospace systems ubiquitously. T represents the total resultant torque which is hoped to equal to the optimal control u^*; $\dot{H}$ represents the change in system angular momentum in the inertial frame; ω represents the angular velocity of the body; J is the inertia matrix for the entire body.

2.2. Luenberger-Like Controllers (i.e., Nonliner-Enhanced Proportional-Derivative-Integral, PDI)

The input torque vector, $[T_x, T_y, T_z]$ is a signal generated by the trajectory block in Figure 1. However, this signal is not tuned to adjust to real world influences, where mechanical hardware can introduce errors due to incorrectly or un-modeled attributes, noise, etc. In order to overcome these losses, either a feed forward controller, a feedback controller, or a combination of both controllers can be used to counter errors. More specifically, proportional, integral, and derivative (PID) gains are correlated only to the position error generated when moving from one position to another position to correct the errors, as displayed in Equation (2). On the other hand, proportional-derivative-integral (PDI) control uses full-state feedback to eliminate virtual-zero references inherent to the cascaded topology of the classical PID form. The modern PDI form is augmented here with a nonlinear term $\omega \times J\omega$ to counter the nonlinear coupling induced by the cross-product operation necessary to account for rigid body motion in moving reference frames, and the combined form is displayed in Equation (3) whose topology is illustrated in Figure 2.

$$Classical\ PID\ feedback\ form:\ u_{fb} = k_p e_\theta + k_d \dot{e}_\theta + k_i \int e_\theta \tag{2}$$

$$Nonlinear-enhanced\ PDI:\ u_{fb} = -k_p(\theta_d - \theta) - k_d(\omega_d - \omega) - k_i \int (\theta_d - \theta)dt - \omega \times J\omega \tag{3}$$

However, using only the position error e_θ, its integral $\int e_\theta dt$, and its derivative $\frac{d}{dt}e_\theta$ results in inaccuracies. This is due in part to noise amplification of the derivative calculation, which is both inefficient and inaccurate as a result of the virtual zero reference created in the cascaded topology of a PID controller. This inaccuracy can be prevented by calculating both the position error and the velocity error, which has been done in this experiment via a nonlinear-enhanced Luenberger proportional, derivative, integral (PDI) controller [15–18] where the enhanced Luenberger Controller differs from the conventional PID controller which only receives a position error and does not have a nonlinear decoupling term to account for moving reference frames. The result is a controller that outputs a commanded torque to the actuator block in Figure 1. Topologies are shown of the overall feedback controller in Figure 4, and the enhanced Luenberger PDI controller in Figure 5. A further augmentation adds a nonlinear component: $\omega \times J\omega$ accounting for coupled motion.

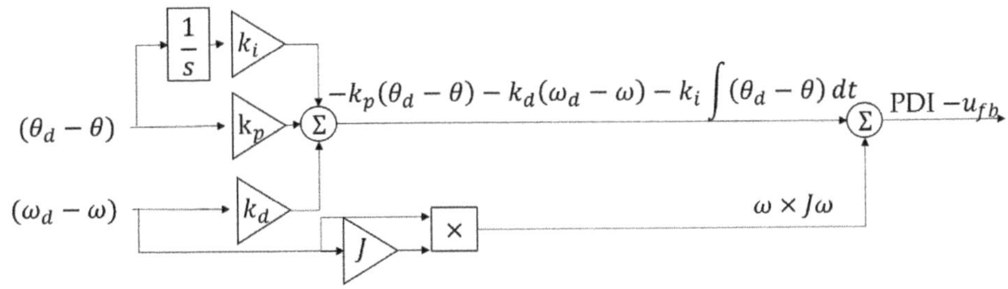

Figure 2. Nonlinear-enhanced PDI controller with desired ω_d input to *remove virtual zero reference* scaled by K_p, K_d, and K_i gains. Also notice the nonlinear cross product $\omega \times J\omega$ enhancement.

2.3. Deterministic Artificial Intelligence

Equation (1) is an analytic governing equation of motion $T = J\dot{\omega} + \omega \times J\omega = u$ substantiates the kinetics in Figure 1 and the nonlinear enhancement in Equation (3). Substituting the desired states ω_d and $\dot{\omega}_d$ into the classical PID control form makes it difficult to see possible exact solutions: $u_{fb} = k_p e_\theta + k_d \dot{e}_\theta + k_i \int e_\theta$, where e_θ is error associated with the angular position state.

Error-Analysis Yields Deterministic Self-Awareness Statement

Equate coefficients of like derivatives of states in Equation (4) and contemplate how they could possibly be analytically equivalent expressions. The comparison is the reason why "???" is placed over the equal sign in the equation. The comparison should lead the reader to replace the equal sign with an unequal sign, since the expression will never be analytically equivalent. The same unfavorable comparison holds true for the nonlinear-enhanced PDI controller, and this inspires the utilization of the exact forms of the governing differential equation of motion in the proposed control as done in Equation (5), the deterministic artificial intelligence self-awareness statement enforcing the rigid body knowledge of its governing equations. Comparison of the methods proposed here will use tracking error analysis in simulations and also computational burden as figures of merit.

$$T = J\dot{\omega} + \omega \times J\omega \overset{???}{\widehat{=}} k_p(\theta_d - \theta) + k_d(\omega_d - \omega) + k_i \int (\theta_d - \theta)dt \tag{4}$$

$$Self\ awareness\ statement\ based\ feedfoward:\ u_{ff} \equiv \hat{J}\dot{\omega}_d + \omega_d \times \hat{J}\omega_d = [\Phi_d]\{\Theta\} \tag{5}$$

$$\Phi_d = \begin{bmatrix} \dot{\omega}_x & \dot{\omega}_y & \dot{\omega}_z & -\omega_y\omega_z & 0 & \omega_z\omega_y \\ \omega_x\omega_z & \dot{\omega}_x & 0 & \dot{\omega}_y & \dot{\omega}_z & -\omega_z\omega_x \\ -\omega_x\omega_y & 0 & \dot{\omega}_x & \omega_y\omega_x & \dot{\omega}_y & \dot{\omega}_z \end{bmatrix}_d \tag{6}$$

$$\Theta = \{J_{xx}, J_{xy}, J_{xz}, J_{yy}, J_{yz}, J_{zz}\}^T \rightarrow \hat{\Theta} = \{\hat{J}_{xx}, \hat{J}_{xy}, \hat{J}_{xz}, \hat{J}_{yy}, \hat{J}_{yz}, \hat{J}_{zz}\}^T \tag{7}$$

Theorem 1. *Deterministic self-awareness statements in a feedforward topology can be optimal (exact) with respect to state tracking error.*

$$T = J\dot{\omega} + \omega \times J\omega = u_{ff} \equiv \hat{J}\dot{\omega}_d + \omega_d \times \hat{J}\omega_d = [\Phi_d]\{\hat{\Theta}\} \tag{8}$$

Proof of Theorem 1. Using the deterministic self-awareness statement in a feedforward topology defines the control as the governing equation of motion: $u_{ff} \equiv \hat{J}\dot{\omega}_d + \omega_d \times \hat{J}\omega_d$.

Corollary stability analysis: As done earlier, equate coefficients of like derivatives of states in Equation (8) and contemplate how they could possibly be analytically equivalent expressions. It is quite easy to see how the deterministic self-awareness statement could lead to analytic solutions. $\forall J \rightarrow \hat{J},\ \dot{\omega} \rightarrow \dot{\omega}_d;\ \omega \rightarrow \omega_d$. □

A control system is capable of learning by estimating the incremental torque error δu using any motion observer and using the estimate to learn the erroneous properties that generated the errors. In a learning control system, the control estimator is the feedforward topology defined by Equation (8) and the corrector or learning mechanism is the feedback defined by Equation (9), where Φ_d and $\hat{\Theta}$ are defined by Equations (6) and (10) (where the * denotes optimality) respectively and where the incremental learning correlating to the incremental error at each time step is $\delta\hat{\Theta}$ in Equation (9).

$$Optimal\ Learning:\ \delta\hat{\Theta}^* = \left([\Phi_d]^T[\Phi_d]\right)^{-1}[\Phi_d]^T\delta u \tag{9}$$

Theorem 2. *Deterministic learning in a feedback topology are optimal with respect to state tracking error.*

$$T = u_{ff} \equiv J\dot{\omega}_d + \omega_d \times J\omega_d = [\Phi_d]\{\Theta\} \rightarrow u^*_{fb} = [\Phi_d]\{\hat{\Theta}^*\} = [\Phi_d]\underbrace{\left([\Phi_d]^T[\Phi_d]\right)^{-1}[\Phi_d]^T\delta u}_{\hat{\Theta}^*} \tag{10}$$

$$\delta u^*_{fb} = [\Phi_d]\{\delta\hat{\Theta}^*\} \rightarrow \{\delta\hat{\Theta}^*\} = [\Phi_d]^{-1}\delta u^*_{fb} = \left(\left(\Phi^T\Phi\right)^{-1}\Phi^T\right)\delta u^*_{fb} \tag{11}$$

Proof of Theorem 2. Error illustrate the control signal to be incorrect, and the control error may be attributed to the mismodeled self-awareness statement by estimating the incremental torque error δu and solving the regression formulated problem for the vector of unknowns: $\hat{\Theta}$. Since the optimal learning expressed in a feedback topology is the 2-norm optimal solution to the standard regression-formulated governing dynamics in Equations (10) and (11), learning is optimal in that sense. □

Corollary stability analysis: When the context of assertion of deterministic self-awareness statements is viewed as a prediction step in a typical Kalman Filter, the subsequent correction-step is stable in accordance with long-held proofs of the stability and optimality of Kalman Filters [51].

Combining Equations (5) and (9) yields a learning system that develops a more accurate time-varying control (12).

$$Optimal\ Learning\ based\ feedback:\ u^*_{fb} = [\Phi_d]\{\hat{\Theta}^*\} = [\Phi_d]\left([\Phi_d]^T[\Phi_d]\right)^{-1}[\Phi_d]^T\delta u \tag{12}$$

$$u_{tot} = u_{ff} + u_{fb}\ \textbf{or}\ u^*_{tot} = u_{ff} + u^*_{fb} \tag{13}$$

$$u^*_{tot} = \Phi\Theta + \Phi\left(\Phi^T\Phi\right)^{-1}\Phi^T du = \Phi\left[\Theta + \left(\Phi^T\Phi\right)^{-1}\Phi^T\delta u\right] \tag{14}$$

Summarizing, replacing the control element of Figure 1 with deterministic artificial intelligence elements of self-awareness statements and learning (as depicted in Figure 3) is accomplished by formulating the term "$[\Phi]\{\Theta\}$" in (8) represents the self-awareness statement written in a regression format with states in the matrix of "knowns" and other variables in the vector of "unknowns". This novel method replaces the feedback control calculation block in Figure 1, and rather than substitute stochastic artificial intelligence in its place, we recommend replacement with self-awareness statements that use feedback to optimally learn. The non-linear state transition matrix $[\Phi]$ was built by knowing the dynamics of the system (embodied in the governing differential equations of rotational mechanics) and $\{\hat{\Theta}\}$ is the estimated vector of unknown variables. Another application includes analyzing a changing inertia matrix, where it is assumed that the mass of the system is varying. The vector of unknowns $\{\hat{\Theta}\}$ is the learned moment of inertia that is recalculated at every iteration of the model and determining its new mass per Equation (9). This analytically exact and now-proven optimal control should immediately (and at every time step) know the precise correct control and is therefore theorized to approach so-called deadbeat control that is correct at the earliest opportunity (without copious training data required by stochastic methods).

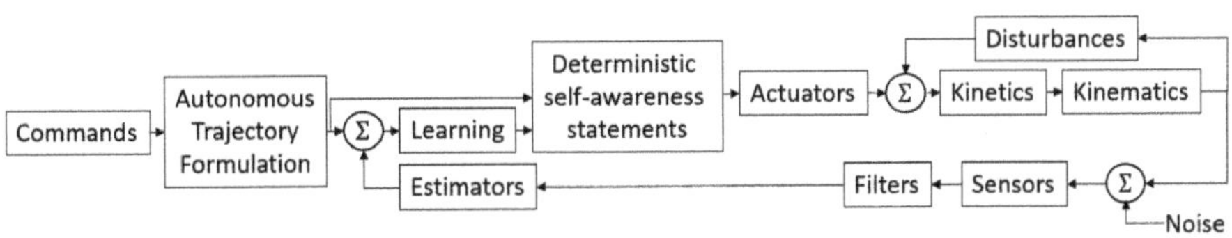

Figure 3. The system topology of deterministic artificial intelligence.

3. Results

This section of the manuscript documents the implementation of three different control algorithm combinations to induce a yawing motion on a rigid body (e.g., autonomous aerospace and maritime systems) by sending the commanded torque signal to a non-redundant array of control moment gyroscope actuators [52]. The three cases investigated are a non-linear feedforward control (case one), a linear feedback control with nonlinear augmentation (case two), and deterministic artificial intelligence as previously presented in a combined non-linear feedforward plus feedback topology of learning (not feedback control) after asserting self-awareness statement. Case one (nonlinear feedforward) implements Equation (5) without the learning, i.e., Equation (7) is not implemented. Case two implements Equation (3) in a typical feedback control topology which is disposed of in case three's implementation of Equations (5) and (7), where optimal learning in Equation (9) supports the implementation of Equation (7). The combined form is Equation (14). The gains for the classical controllers used for comparison are found in Table 1. Notice that Table 1 includes the gains from the Luenberger observer used to find the control-error δu necessary for optimal learning. It will be seen that replacement of typical feedforward and feedback topologies with deterministic A.I. will be the most accurate option simultaneously achieving the lowest computational burden.

Table 1. Tuned gain values [1] for the PDI controller and observer.

	k_p Gain	k_d Gain	k_i Gain
Enhanced-PDI Controller	1000	10	0.1
Luenberger Observer for δu	1000	10	0.1

[1] Gains fixed to highlight topological differences.

The model of the rigid body in this manuscript was built in Matlab and Simulink, where integrations were calculated using the ode45 with Runge-Kutta solver with a fixed time-step. Euler Angles were resolved using a 3-2-1 rotation sequence with the atan2 trigonometry function to eliminate quadrant ambiguity.

Utilizing the same rigid body model and sinusoidal trajectory generation as reference [1], initialized values include: torque = [0, 0, 0] and quaternion = [0, 0, 0, 1]. The rigid body's inertia matrix is $J = [10, 0.1, 0.1; 0.1, 10, 0.1; 0.1, 0.1, 10]$. The realistic disturbance torques are defined in reference [5]. The orbital altitude was set at 150 km with an atmospheric drag coefficient of 2.5. Each simulation utilized a five second quiescent period to validate the model, five second maneuver time, and five second post maneuver observation period, totaling 15 s for a large angle slew.

3.1. Time-Step Analysis

Time-step analysis was completed to determine whether reducing the time-step would help minimize the error deviations between the body frame and the inertial frame. The results of executing a maneuver with deterministic artificial intelligence and two different time-steps is depicted in Figure 4. Expectations were that a smaller time-step would result in more precise results, meaning a smaller deviation between the commanded and executed Euler angles. However, comparing the trajectories within each of the three sub-plots in Figure 4 shows that although some refinement is gained by decreasing the time-step, the gain is minimal. Therefore, a larger time-step (e.g., 0.01 s) can be used with slight degradation.

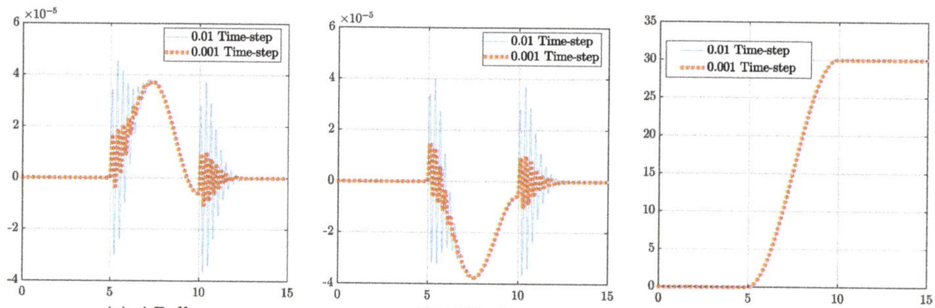

(**a**) ϕ Roll (degrees versus seconds) (**b**) θ Pitch (degrees versus seconds) (**c**) ψ Yaw (degrees versus seconds)

Figure 4. Time-step analysis for the ϕ, θ, and ψ Euler angles for two disparate time-steps with deterministic artificial intelligence. Pay particular attention to the near coincident performances when viewed in large-scale in Figure 4c, while the smaller-scaled plots reveal differences.

Comparing the $\theta_{desired} - \theta_{actual}$ and $\omega_{desired} - \omega_{actual}$ errors for time-steps of 0.01 and 0.001 in Figure 5 yielded a similar result. Therefore, these results confirm that varying the time-step has limited impact on the trajectories. With this knowledge, for the gains in Table 1, a minimum time-step of 0.01 is recommended, since decreasing step-size any further provided slight benefit.

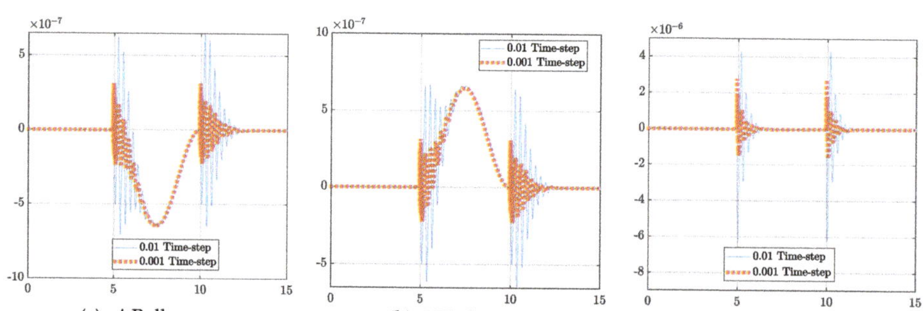

(**a**) ϕ Roll error (degrees versus seconds) (**b**) θ Pitch error (degrees versus seconds) (**c**) ψ Yaw error (degrees versus seconds)

Figure 5. *Cont.*

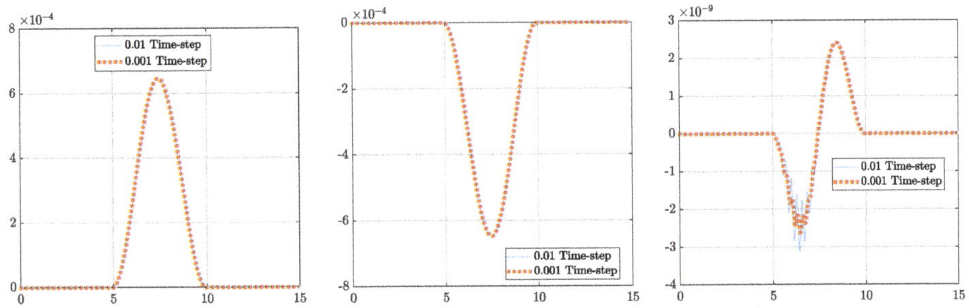

(**d**) ω_x Roll rate (degrees/second versus seconds)
(**e**) ω_y Pitch rate (degrees/second versus seconds)
(**f**) ω_z Yaw rate (degrees/second versus seconds)

Figure 5. Time-step analysis comparing $\theta_{actual} - \theta_{desired}$ and $\omega_{actual} - \omega_{desired}$ errors.

3.2. Control Implementation

The performance of the three control system implementations is depicted in Figure 6. Comparing the three cases allowed further analysis on the differences between feedforward, feedback, and the combined deterministic artificial intelligence system. The feedforward and the deterministic artificial intelligence systems are more precise than the feedback method. This is because they are based on exact control Equations (9) and (10) respectively. Conversely, the feedback controller is based off a PDI controller that suffers phase lag [53], and is therefore less precise. Additionally, the gains in a PDI or PID controller must be finely tuned with pre-determined gain values, which can be an iterative and time-consuming task because controller performance varies greatly depending on the values. Lastly, the deterministic artificial intelligence configuration represents an error combination of both the feedforward and feedback plots. This allows the analytical accuracy of the feedforward equation to be updated with the responsiveness of the feedback correction by learning. Plot scaling precludes easy visualization of the error, so tabulated numerical results are provided in Table 2.

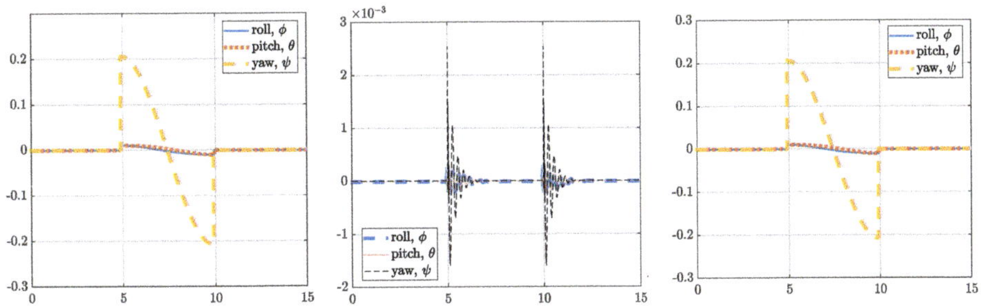

(**a**) Nonlinear feedforward
(**b**) Linear feedback
(**c**) Deterministic artificial intelligence

Figure 6. Control (Newton-meters) versus time in seconds for the three configurations.

Table 2. Actual desired Euler angle errors and associated run times for the three cases [1].

	$\phi_{actual} - \phi_{desired}$ **(degrees)**	$\theta_{actual} - \theta_{desired}$ **(degrees)**	$\psi_{actual} - \psi_{desired}$ **(degrees)**	**Computational Burden (s)**
u_{ff}	-3.93×10^{-4}	4.18×10^{-4}	-1.48×10^{-2}	24.7
u_{fb}	2.30×10^{-8}	$-7.31\text{e} \times 10^{-8}$	2.97×10^{-8}	36.3
D.A.I.	1.10×10^{-8}	-7.55×10^{-9}	4.85×10^{-9}	24.1

[1] using a 0.001 time-step. Reminder: ψ is the axis of maneuver.

Figure 7 shows the Euler angle tracking error over time. Figure 7a,b shows that the error is different for each controller. The feedback controller fluctuates initially as it corrects to reduce error over time. The feedforward controller is excellent initially, but slowly deviates as error accrues without correction. Lastly, the deterministic artificial intelligence system is the best, starting with minimal error, and furthermore correcting that error over time.

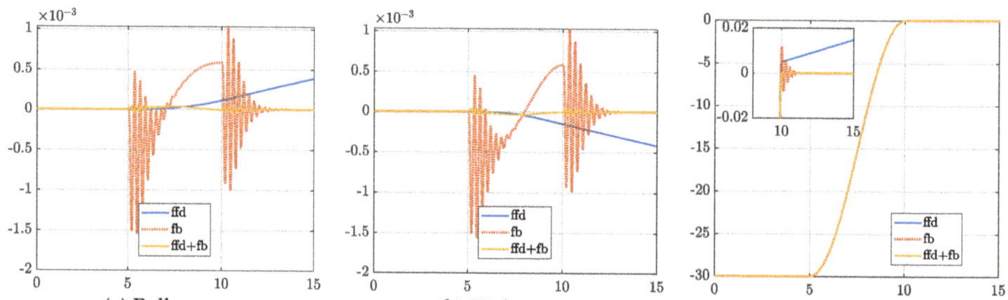

(**a**) Roll error (degrees versus seconds) (**b**) Pitch error (degrees versus seconds) (**c**) Yaw error (degrees versus seconds)

Figure 7. Euler Angle error for the three controller configurations.

Figure 7c shows the position error in the yaw ψ channel, which is the channel in which a 30-degree maneuver was commanded. The feedforward controller again starts off with minimal error before deviating over time, while the feedback controller starts with the greatest amount of error that is quickly damped, and the deterministic artificial intelligence is the best of both.

Table 2 compares the boundary condition satisfaction at the final time of the maneuver. The results show that the deterministic artificial intelligence system (D.A.I.) has both the least amount of error and the shortest computational runtime. The feedforward controller is the worst in accuracy due to an inability to correct for error, while the feedback controller can correct, but takes longer to do so. The feedback controller is hypothesized to perform worse because Equation (2) tries to model Equation (1), but can only poorly approximate it, yielding inaccuracies.

Figure 8 is a revisualization of the data presented in Figure 7 for clarity of conclusions. This depiction is more intuitive and illustrates the change in angular yaw position over time for each controller, as well as magnifying the post maneuver oscillations and damping (or lack of such, in the case of Figure 8a,b). Commanding $\{\ \phi(t_f)\ \ \theta(t_f)\ \ \psi(t_f)\ \}^T = \{\ 0\ \ 0\ \ 30\ \}^T$, we see that all controller configurations are responsive to this input, with expected differences. The accuracy of the feedforward control in Figure 8a, combined with the lightly damped response of feedback control in Figure 8b, clearly illustrate the superior, dampened response of the deterministic artificial intelligence controller in Figure 8c.

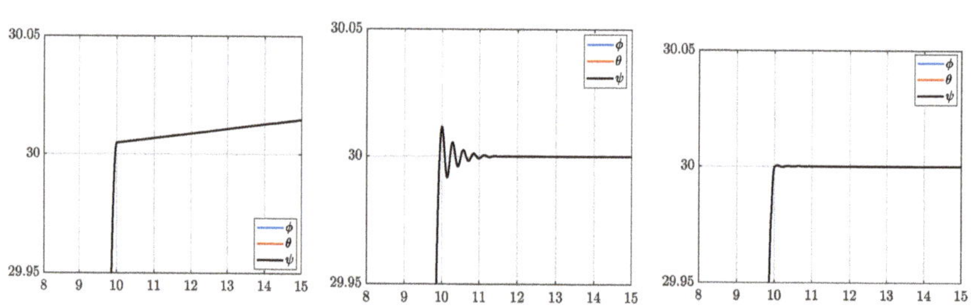

(**a**) Zoomed yaw error (degrees versus seconds) nonlinear feedfoward (**b**) Zoomed yaw error (degrees versus seconds) classical feedback (**c**) Zoomed yaw error (degrees versus seconds) deterministic artificial intelligence

Figure 8. Change in angular position for all three controller configurations. (degrees versus seconds).

4. Discussion

The implemented experiment compared the effects of a feedforward, feedback, and a deterministic artificial intelligence system. A yaw maneuver was commanded, and the response measured to show that an attitude determination and control system can estimate and then update its control over time, akin to feedforward and feedback (where feedback here is a learning mechanism). The results showed that the feedforward controller lacks a correction mechanism that accrues error, the feedback system requires more time to correct for error that starts in the system, and the deterministic artificial intelligence system combines the best of both systems for superior accuracy in the lower computational burden. Therefore, the deterministic artificial intelligence system is the best choice for its accuracy and adaptability. However, this combined system needs to be further researched by subjecting the system to noise and induced parameter variation (including disturbances) to validate the system's responsiveness. Future research will implement time-variations in the rigid body's parameters to simulate damage from inelastic collisions, or alternatively highly energetic elastic collisions that remove significant part of the rigid body. It is theorized that deterministic artificial intelligence could quickly (and optimally) recover.

Readers seeking apply the methods proposed in this manuscript may follow the procedures outlined in the flowchart in Figure 9. System definition acknowledges the first principle governing the rigid body of interest, e.g., spacecraft, aircraft, underwater vehicles, etc.

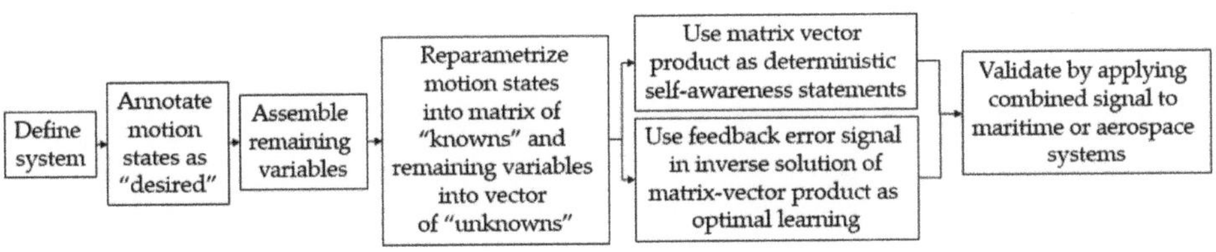

Figure 9. Technology development flowchart using the proposed methods.

Future Research

Although the methods proposed here stem from analysis of the errors between the algorithms and the systems being controlled, typical stability analysis for nonlinear learning systems would often include phase portraits and also analytic demonstrations using Lyapunov's first and second methods. While these were unnecessary here, it remains for future researchers to validate stability demonstrated here with either of these three methods.

Author Contributions: Conceptualization, B.S., A.R., and T.S.; Methodology, T.S.; Software, B.S., A.R., and T.S.; Validation, B.S., A.R., and T.S.; Formal analysis, B.S., A.R., and T.S.; Investigation, B.S., A.R., and T.S.; Resources, T.S.; Data curation, B.S., A.R., and T.S.; Writing—original draft preparation, B.S. and A.R.; Writing—review and editing, B.S., A.R., and T.S.; Visualization, B.S., A.R., and T.S.; Supervision, T.S.; Project administration, B.S. and T.S.; Funding acquisition, T.S. All authors have read and agreed to the published version of the manuscript.

Funding: This research received no external funding.

Conflicts of Interest: The authors declare no conflict of interest.

References

1. Smeresky, B.; Rizzo, A.; Sands, T. Kinematics in the Information Age. *Mathematics* **2018**, *6*, 148. [CrossRef]
2. Cooper, M.; Heidlauf, P.; Sands, T. Controlling Chaos—Forced van der pol equation. *Mathematics* **2017**, *5*, 70. [CrossRef]
3. Sands, T. Nonlinear-Adaptive Mathematical System Identification. *Computation* **2017**, *5*, 47. [CrossRef]
4. Nakatani, S.; Sands, T. Battle-damage tolerant automatic controls. *Electr. Electron. Eng.* **2018**, *8*, 10–23.
5. Nakatani, S.; Sands, T. Simulation of rigid body damage tolerance and adaptive controls. In Proceedings of the IEEE Aerospace Conference, Big Sky, MT, USA, 1–8 March 2014; pp. 1–16.

6. Nakatani, S.; Sands, T. Autonomous Damage Recovery in Space. *Int. J. Autom. Control Intell. Syst.* **2016**, *2*, 23–36.
7. EPICA Lastest News Blog. Deterministic Approach to Machine Learning and AI. 2018. Available online: https://www.epica.ai/thinking/blog/deterministic-approach-to-machine-learning-and-ai.html (accessed on 19 November 2019).
8. Deterministic Data-based AI Key for Security. Available online: https://www.computerweekly.com/news/252464484/Deterministic-data-based-AI-key-for-security (accessed on 25 November 2019).
9. DeGrazia, D. Self-awareness in animals. In *The Philosophy of Animal Minds*; Lurz, R.W., Ed.; Cambridge University Press: Cambridge, UK, 2009; pp. 201–217.
10. Bekoff, M. Awareness: Animal reflections. *Nature* **2002**, *419*, 255. [CrossRef]
11. Gallup, G.G., Jr.; Anderson, J.R.; Shillito, D.J. The mirror test. In *The Cognitive Animal: Empirical and Theoretical Perspectives on Animal Cognition*; Bekoff, M., Allen, C., Burghardt, G.M., Eds.; MIT Press: Cambridge, MA, USA, 2002; pp. 325–333.
12. Plotnik, J.M.; De Waal, F.B.; Reiss, D. Self-recognition in an Asian elephant. *Proc. Natl. Acad. Sci. USA* **2006**, *103*, 17053–17057. [CrossRef]
13. Prior, H.; Schwarz, A.; Güntürkün, O. Mirror-Induced Behavior in the Magpie (Pica pica): Evidence of Self-Recognition. *PLoS Biol.* **2008**, *6*, e202. [CrossRef]
14. Alison, M. Mirror Test Shows Magpies Aren't So Bird-Brained. Available online: https://www.newscientist.com/article/dn14552-mirror-test-shows-magpies-arentso-birdbrained.html#.VHVdHf4tDIV (accessed on 26 November 2014).
15. Tennesen, M. Do Dolphins Have a Sense of Self? *Natl. Wildl.* **2003**, *41*, 66.
16. Bard, K. Self-Awareness in Human and Chimpanzee Infants: What is Measured and What is Meant by the Mark and Mirror Test? *Infancy* **2006**, *9*, 191–219. [CrossRef]
17. Cammaerts Tricot, M.C.; Cammaerts, R. Are Ants (Hymenoptera Formicide) Capable of Self Recognition? *J. Sci.* **2015**, *5*, 521–532.
18. Pfeifer, R.; Bongard, J. *How the Body Shapes the Way We Think: A New View of Intelligence*; MIT Press: Cambridge, MA, USA, 2006.
19. Lieberman, P. *The Unpredictable Species: What Makes Humans Unique*; Princeton University Press: Princeton, NJ, USA, 2013.
20. Rochat, P. Five levels of self-awareness as they unfold early in life. *Conscious. Cognit.* **2003**, *12*, 717–731. [CrossRef]
21. Geangu, E. Notes on Self-Awareness Development in Early Infancy. *Cognit. Brain Behav.* **2008**, *12*, 103–113.
22. Yawkey, T.D.; Johnson, J.E. (Eds.) *Integrative Processes and Socialization Early to Middle Childhood*; Psychology Press: East Sussex, UK, 2013; ISBN 9780203767696.
23. Rochat, P. Self-Perception and Action in Infancy. *Exp. Brain Res.* **1998**, *123*, 102–109. [CrossRef]
24. Broesch, T.; Callaghan, T.; Henrich, J.; Murphy, C.; Rochat, P. Cultural Variations in Children's Mirror Self-Recognition. *J. Cross Cult. Psychol.* **2011**, *42*, 1018–1029. [CrossRef]
25. Locke, J. *An Essay Concerning Human Understanding*; Wentworth Press: Sidney, Australia, 1775; ISBN 0526315180.
26. Kendra, C. What Is Self-Awareness? Available online: http://psychology.about.com/od/cognitivepsychology/fl/What-Is-Self-Awareness.htm (accessed on 25 November 2014).
27. Abraham, S. *Goldstein the Insanity Defense*; Yale University Press: New Haven, CT, USA, 1967; p. 9. ISBN 978-0-300-00099-3.
28. Uddin, L.Q.; Davies, M.S.; Scott, A.A.; Zaidel, E.; Bookheimer, S.Y.; Iacoboni, M.; Dapretto, M. Neural Basis of Self and Other Representation in Autism: An fMRI Study of Self-Face Recognition. *PLoS ONE* **2008**, *3*, e3526. [CrossRef]
29. Razzaq, M.A.; Villalonga, C.; Lee, S.; Akhtar, U.; Ali, M.; Kim, E.-S.; Khattak, A.M.; Seung, H.; Hur, T.; Bang, J.; et al. mlCAF: Multi-Level Cross-Domain Semantic Context Fusioning for Behavior Identification. *Sensors* **2017**, *17*, 2433. [CrossRef]
30. Sysoev, M.; Kos, A.; Guna, J.; Pogačnik, M. Estimation of the Driving Style Based on the Users' Activity and Environment Influence. *Sensors* **2017**, *17*, 2404. [CrossRef]

31. Roldán, J.J.; Peña-Tapia, E.; Martín-Barrio, A.; Olivares-Méndez, M.A.; Del Cerro, J.; Barrientos, A. Multi-Robot Interfaces and Operator Situational Awareness: Study of the Impact of Immersion and Prediction. *Sensors* **2017**, *17*, 1720. [CrossRef]
32. Mcheick, H.; Saleh, L.; Ajami, H.; Mili, H. Context Relevant Prediction Model for COPD Domain Using Bayesian Belief Network. *Sensors* **2017**, *17*, 1486. [CrossRef]
33. García, Ó.; Alonso, R.S.; Prieto, J.; Corchado, J.M. Energy Efficiency in Public Buildings through Context-Aware Social Computing. *Sensors* **2017**, *17*, 826. [CrossRef] [PubMed]
34. Chang, J.; Yao, W.; Li, X. A Context-Aware S-Health Service System for Drivers. *Sensors* **2017**, *17*, 609. [CrossRef] [PubMed]
35. Scholze, S.; Barata, J.; Stokic, D. Holistic Context-Sensitivity for Run-Time Optimization of Flexible Manufacturing Systems. *Sensors* **2017**, *17*, 455. [CrossRef]
36. Moore, P. Do We Understand the Relationship between Affective Computing, Emotion and Context-Awareness? *Machines* **2017**, *5*, 16. [CrossRef]
37. Lewis, P.R.; Platzner, M.; Rinner, B.; Tørresen, J.; Yao, X. (Eds.) *Self-Aware Computing Systems: An Engineering Approach*; Springer International Publishing AG: Cham, Switzerland, 2016.
38. Abba, S.; Lee, J.-A. An Autonomous Self-Aware and Adaptive Fault Tolerant Routing Technique for Wireless Sensor Networks. *Sensors* **2015**, *15*, 20316–20354. [CrossRef]
39. Kao, H.-A.; Jin, W.; Siegel, D.; Lee, J. A Cyber Physical Interface for Automation Systems—Methodology and Examples. *Machines* **2015**, *3*, 93–106. [CrossRef]
40. Jauk, E.; Kanske, P. Perspective Change and Personality State Variability: An Argument for the Role of Self-Awareness and an Outlook on Bidirectionality (Commentary on Wundrack et al., 2018). *J. Intell.* **2019**, *7*, 10. [CrossRef]
41. Wundrack, R.; Prager, J.; Asselmann, E.; O'Connell, G.; Specht, J. Does Intraindividual Variability of Personality States Improve Perspective Taking? An Ecological Approach Integrating Personality and Social Cognition. *J. Intell.* **2018**, *6*, 50. [CrossRef]
42. Kosak, O.; Wanninger, C.; Hoffmann, A.; Ponsar, H.; Reif, W. Multipotent Systems: Combining Planning, Self-Organization, and Reconfiguration in Modular Robot Ensembles. *Sensors* **2019**, *19*, 17. [CrossRef]
43. Van Pham, H.; Moore, P. Robot Coverage Path Planning under Uncertainty Using Knowledge Inference and Hedge Algebras. *Machines* **2018**, *6*, 46. [CrossRef]
44. Sands, T.; Kim, J.J.; Agrawal, B.N. Spacecraft fine tracking pointing using adaptive control. In Proceedings of the 58th International Astronautical Congress, Hyderabad, India, 24–28 September 2007.
45. Sands, T.; Kim, J.J.; Agrawal, B. Spacecraft Adaptive Control Evaluation. In Proceedings of the Infotech@Aerospace, Garden Grove, CA, USA, 19–21 June 2012.
46. Sands, T.; Kim, J.J.; Agrawal, B. Improved Hamiltonian adaptive control of spacecraft. In Proceedings of the Aerospace Conference, Big Sky, MT, USA, 7–14 March 2009; pp. 1–10.
47. Sands, T.; Lorenz, R. Physics-Based Automated Control of Spacecraft. In Proceedings of the AIAA Space Conference & Exposition, Pasadena, CA, USA, 14–17 September 2009.
48. Sands, T. Physics-Based Control Methods. In *Advances in Spacecraft Systems and Orbit Determination*; InTech Publishers: London, UK, 2012; pp. 29–54.
49. Sands, T. Improved Magnetic Levitation via Online Disturbance Decoupling. *Phys. J.* **2015**, *1*, 272–280.
50. Baker, K.; Cooper, M.; Heidlauf, P.; Sands, T. Autonomous trajectory generation for deterministic artificial intelligence. *Electr. Electron. Eng.* **2018**, *8*, 59–68.
51. Kalman, R.E. A New Approach to Linear Filtering and Prediction Problems. *J. Basic Eng.* **1960**, *82*, 35. [CrossRef]
52. Sands, T. Control Moment Gyroscope Singularity Reduction via Decoupled Control. In Proceedings of the IEEE SEC Proceedings, Atlanta, GA, USA, 5–8 March 2009.
53. Sands, T. Phase Lag Elimination at All Frequencies for Full State Estimation of Rigid body Attitude. *Phys. J.* **2017**, *3*, 1–12.

algorithms

MDPI

Article

Learning Output Reference Model Tracking for Higher-Order Nonlinear Systems with Unknown Dynamics †

Mircea-Bogdan Radac * and Timotei Lala

Department of Automation and Applied Informatics, Politehnica University of Timisoara, 2 Bd. V. Parvan, 300223 Timisoara, Romania; timotei.lala@student.upt.ro

* Correspondence: bogdan.ttl@gmail.com; Tel.: +40-256-403240; Fax: +40-256-403214

† This paper is an extended version of our paper published in the 27th Mediterranean Conference on Control and Automation (MED 2019).

Received: 1 May 2019; Accepted: 9 June 2019; Published: 12 June 2019

Abstract: This work suggests a solution for the output reference model (ORM) tracking control problem, based on approximate dynamic programming. General nonlinear systems are included in a control system (CS) and subjected to state feedback. By linear ORM selection, indirect CS feedback linearization is obtained, leading to favorable linear behavior of the CS. The Value Iteration (VI) algorithm ensures model-free nonlinear state feedback controller learning, without relying on the process dynamics. From linear to nonlinear parameterizations, a reliable approximate VI implementation in continuous state-action spaces depends on several key parameters such as problem dimension, exploration of the state-action space, the state-transitions dataset size, and a suitable selection of the function approximators. Herein, we find that, given a transition sample dataset and a general linear parameterization of the Q-function, the ORM tracking performance obtained with an approximate VI scheme can reach the performance level of a more general implementation using neural networks (NNs). Although the NN-based implementation takes more time to learn due to its higher complexity (more parameters), it is less sensitive to exploration settings, number of transition samples, and to the selected hyper-parameters, hence it is recommending as the de facto practical implementation. Contributions of this work include the following: VI convergence is guaranteed under general function approximators; a case study for a low-order linear system in order to generalize the more complex ORM tracking validation on a real-world nonlinear multivariable aerodynamic process; comparisons with an offline deep deterministic policy gradient solution; implementation details and further discussions on the obtained results.

Keywords: approximate dynamic programming; reinforcement learning; data-driven control; model-free control; reference trajectory tracking; output reference model; multivariable control; aerodynamic rotor system; neural networks; learning systems

1. Introduction

The output reference model (ORM) tracking problem is of significant interest in practice, especially for nonlinear systems control, since by selection of a linear ORM, feedback linearization is enforced on the controlled process. Then, the closed-loop control system can act linearly in a wide range and not only in the vicinity of an operating point. Subsequently, linearized control systems are then subjected to higher level learning schemes such as the Iterative Learning Control ones, with practical implications such as primitive-based learning [1] that can extrapolate optimal behavior to previously unseen tracking scenarios.

On another side, selection of a suitable ORM is not straightforward because of several reasons. The ORM has to be matched with the process bandwidth and with several process nonlinearities such as, e.g., input and output saturations. From classical control theory, dead-time and non-minimum-phase characters of the process cannot be compensated for and must be reflected in the ORM. Apart from this information that can be measured or inferred from working experience with the process, avoiding knowledge of the process' state transition function (process dynamics)—the most time consuming to identify and the most uncertain part of the process—in designing high performance control is very attractive in practice.

Reinforcement Learning (RL) has developed both from the artificial intelligence [2], and from classical control [3–7], where it is better known as Adaptive (Approximate, Neuro) Dynamic Programming (ADP). Certain ADP variants can ensure ORM tracking control without knowing the state-space (transition function) dynamics of the controlled process, which is of high importance in the practice of model-free (herein accepted as unknown dynamics) and data-driven control schemes that are able to compensate for poor modeling and process model uncertainty. Thus, ADP relies only on data collected from the process called state transitions. While plenty of mature ADP schemes already exist in the literature, tuning such schemes for a particular problem requires significant experience. Firstly, it must be specified whether ADP deals with continuous (infinite) or discrete (finite) state-action spaces. Then, the intended implementation will decide upon online/offline and/or adaptive/batch processing, the suitable selection of the approximator used for the extended cost function (called the Q-function) and/or for the controller. Afterwards, linear or nonlinear parameterizations are sought. Exploration of the state-action spaces is critical, as well as the hyperparameters of the overall learning scheme such as the number of transition samples, trading off exploration with exploitation, etc. Although successful stories on RL and ADP applied to large state-action spaces are reported mainly with artificial intelligence [8], in control theory, most approaches use low-order processes as representative case studies and mainly in linear quadratic regulator (LQR)-like settings (regulating states to zero). While, in an ADP, the reference input tracking control problem has been tackled before for linear time-invariant (LTI) processes by the name of Linear Quadratic Tracking (LQT) [9,10], the ORM tracking for nonlinear processes was rarely addressed [11,12].

The iterative model-free approximate Value Iteration (IMF-AVI) proposed in this work belongs to the family of batch-fitted Q-learning schemes [13,14] known as action-dependent heuristic dynamic programming (ADHDP) that are popular and representative ADP approaches, owing to their simplicity and model-free character. These schemes have been implemented in many variants: online vs. offline, adaptive or batch, for discrete/continuous states and actions, with/without function approximators, such as Neural Networks (NNs) [12,15–23].

Concerning the exploration issue in ADP for control, a suitable exploration that covers as well as possible the state-action space is not trivially ensured. Randomly generated control input signals will almost surely fail to guide the exploration in the entire state-action space, at least not in a reasonable amount of time. Then, a priori designed feedback controllers can be used under a variable reference input serving to guide the exploration [24]. The existence of an initial feedback stabilizing controller, not necessarily of a high performance one, can accelerate the transition samples dataset collection under exploration. This allows for offline IMF-AVI based on large datasets, leading to improved convergence speed for high-dimensional processes. However, such input–output (IO) or input-state feedback controllers were traditionally not to be designed without using a process model, until the advent of data-driven model-free controller design techniques that have appeared from the field of control theory: Virtual Reference Feedback Tuning (VRFT) [25], Iterative Feedback Tuning [26], Model Free Iterative Learning Control [27–29], Model Free (Adaptive) Control [30,31], with representative applications [32–34]. This work shows a successful example of a model-free output feedback controller used to collect input-to-state transition samples from the process for learning state-feedback ADP-based ORM tracking control. Therefore it fits with the recent data-driven control [35–43] and reinforcement learning [12,44,45] applications.

The case study deals with the challenging ORM tracking control for a nonlinear real-world two-inputs–two-outputs aerodynamic system (TITOAS) having six natural states that are extended with four additional ones according to the proposed theory. The process uses aerodynamic thrust to create vertical (pitch) and horizontal (azimuth) motion. It is shown that IMF-AVI can be used to attain ORM tracking of first order lag type, despite the high order of the multivariable process, and despite the pitch motion being naturally oscillatory and the azimuth motion practically behaving close to an integrator. The state transitions dataset is collected under the guidance of an input–output (IO) feedback controller designed using model-free VRFT (© 2019 IEEE [12]).

As a main contribution, the paper is focused on a detailed comparison of the advantages and disadvantages of using linear and nonlinear parameterizations for the IMF-AVI scheme, while covering complete implementation details. To the best of authors' knowledge, the ORM tracking context with linear parameterizations was not studied before for high-order real-world processes. Moreover, theoretical analysis shows convergence of the IMF-AVI while accounting for approximation errors and explains for the robust learning convergence of the NN-based IMF-AVI. The results indicate that the nonlinearly parameterized NN-based IMF-AVI implementation should be *de facto* in practice since, although more time-consuming, it automatically manages the basis function selection, it is more robust to dataset size and exploration settings, and generally more well-suited for nonlinear processes with unknown dynamics. The main updates with respect to our paper [12] include: detailed IMF-AVI convergence proofs under general function approximators; a case study for a low order linear system in order to generalize to the more complex ORM tracking validation on the TITOAS process; comparisons with an offline Deep Deterministic Policy Gradient solution; more implementation details and further discussions on the obtained results.

Section 2 is dedicated to the formalization of the ORM tracking control problem, while Section 3 proposes a solution to this problem using an IMF-AVI approach. Section 4 validates the proposed approach on the TITOAS system, with concluding remarks presented in Section 5.

2. Output Model Reference Control for Unknown Dynamics Nonlinear Processes

2.1. The Process

A discrete-time nonlinear unknown open-loop stable state-space deterministic strictly causal process is defined as [12,46]

$$P : \{\mathbf{x}_{k+1} = \mathbf{f}(\mathbf{x}_k, \mathbf{u}_k), \mathbf{y}_k = \mathbf{g}(\mathbf{x}_k)\}, \tag{1}$$

where k indexes the discrete time, $\mathbf{x}_k = [x_{k,1}, ..., x_{k,n}]^\top \in \Omega_X \subset \mathbb{R}^n$ is the n-dimensional state vector, $\mathbf{u}_k = [u_{k,1}, ..., u_{k,m_u}]^\top \in \Omega_U \subset \mathbb{R}^{m_u}$ is the control input signal, $\mathbf{y}_k = [y_{k,1}, ..., y_{k,p}]^\top \in \Omega_Y \subset \mathbb{R}^p$ is the measurable controlled output, $\mathbf{f} : \Omega_X \times \Omega_U \mapsto \Omega_X$ is an *unknown* nonlinear system function continuously differentiable within its domain, $\mathbf{g} : \Omega_X \mapsto \Omega_Y$ is an unknown nonlinear continuously differentiable output function. Initial conditions are not accounted for at this point. Assume known domains $\Omega_X, \Omega_U, \Omega_Y$ are compact convex. Equation (1) is a general un-restrictive form for most controlled processes. The following assumptions common to the data-driven formulation are [12,46]:

Assumption 1 (A1). *(1) is fully state controllable with measurable states.*

Assumption 2 (A2). *(1) is input-to-state stable on known domain $\Omega_U \times \Omega_X$.*

Assumption 3 (A3). *(1) is minimum-phase (MP).*

A1 and A2 are widely used in data-driven control, cannot be checked analytically for the unknown model (1) but can be inferred from historical and working knowledge with the process. Should such information not be available, the user can attempt process control under restraining safety operating conditions, that are usually dealt with at supervisory level control. Input to state stability (A2) is necessary if open-loop input-state samples collection is intended to be used for state space control design. Assumption A2 can be relaxed if a stabilizing state-space controller is already available and used just for the purpose of input-state data collection. A3 is the least restrictive assumption and it is used in the context of the VRFT design of a feedback controller based on input–output process data. Although solutions exist to deal with nonminimum-phase systems processes, the MP assumption simplifies the VRFT design and the output reference model selection (to be introduced in the following section).

Comment 1. [46] Model (1) accounts for a wide range of processes including fixed time-delay ones. For positive integer nonzero delay d on the control input $\mathbf{u}_{k-d}$, additional states can extend the initial process model (1) as $\mathbf{x}^u_{k,1} = \mathbf{u}_{k-1}, \mathbf{x}^u_{k,2} = \mathbf{u}_{k-2}, ..., \mathbf{x}^u_{k,d} = \mathbf{u}_{k-d}$ and arrive at a state-space model without delays, in which the additional states are measurable as past input samples. A delay in the original states in (1), i.e., $\mathbf{x}_{k-d}$, are similarly treated.

2.2. Output Reference Model Control Problem Definition

Let the discrete-time known open-loop stable minimum-phase (MP) state-space deterministic strictly causal ORM be [12,46]

$$M : \{\mathbf{x}^m_{k+1} = \mathbf{f}^m(\mathbf{x}^m_k, \mathbf{r}_k), \mathbf{y}^m_k = \mathbf{g}^m(\mathbf{x}^m_k)\}, \tag{2}$$

where $\mathbf{x}^m_k = [x^m_{k,1}, ..., x^m_{k,n}]^\top \in \Omega_{X_m} \subset \mathbb{R}^n_m$ is the ORM state, $\mathbf{r}_k = [r_{k,1}, ..., r_{k,p}]^\top \in \Omega_{R_m} \subset \mathbb{R}^p$ is the reference input signal, $\mathbf{y}^m_k = [y^m_{k,1}, ..., y^m_{k,p}]^\top \in \Omega_{Y_m} \subset \mathbb{R}^p$ is the ORM output, $\mathbf{f}^m : \Omega_{X_m} \times \Omega_{R_m} \mapsto \Omega_{X_m}$, $\mathbf{g}^m : \Omega_{X_m} \mapsto \Omega_{Y_m}$ are *known* nonlinear mappings. Initial conditions are zero unless otherwise stated. Notice that $\mathbf{r}_m, \mathbf{y}_k, \mathbf{y}^m_k$ are size p for square feedback control systems (CSs). If the ORM (2) is LTI, it is always possible to express the ORM as an IO LTI transfer function (t.f.) $\mathbf{M}(z)$ ensuring $\mathbf{y}^m_k = \mathbf{M}(z)\mathbf{r}_k$, where $\mathbf{M}(z)$ is commonly an asymptotically stable unit-gain rational t.f. and $\mathbf{r}_k$ is the reference input that drives both the feedback CS and the ORM. We introduce an extended process comprising of the process (1) coupled with the ORM (2). For this, we consider the reference input $\mathbf{r}_k$ as a set of measurable exogenous signals (possibly interpreted as a disturbance) that evolve according to $\mathbf{r}_{k+1} = \mathbf{h}^m(\mathbf{r}_k)$, with known nonlinear $\mathbf{h}^m : \mathbb{R}^m \mapsto \mathbb{R}^m$, where $\mathbf{r}_k$ is measurable. Herein, $\mathbf{h}^m(.)$ is a generative model for the reference input (© 2019 IEEE [12]).

The class of LTI generative models $\mathbf{h}^m(.)$ has been studied before in [9] but it is a rather restrictive one. For example, reference inputs signals modeled as a sequence of steps of constant amplitude cannot be modeled by LTI generative models. A step reference input signal with constant amplitude over time can be modeled as $\mathbf{r}_{k+1} = \mathbf{r}_k$ with some initial condition $\mathbf{r}_0$. On the other hand, a sinusoidal scalar reference input signal r_k can be modeled only through a second order state-space model. To see this, let the Laplace transform of $cos(\omega t)\sigma(t)$ ($\sigma(t)$ is the unit step function) be $H(s) = \ell\{cos(\omega t)\sigma(t)\}$ with the complex Laplace variable s. If $sH(s)$ is considered a t.f. driven by the unit step function with Laplace transform $\ell\{\sigma(t)\} = 1/s$, then the LTI discrete-time state-space associated with $sH(s)$ acting as a generative model for r_k is of the form

$$\begin{aligned} \mathbf{o}_{k+1} &= \mathbf{A}\mathbf{o}_k + \mathbf{B}\sigma_k, \\ r_k &= \mathbf{C}\mathbf{o}_k + \mathbf{D}\sigma_k, \end{aligned} \tag{3}$$

with known $\mathbf{A} \in \mathbb{R}^{2\times 2}, \mathbf{B} \in \mathbb{R}^{2\times 1}, \mathbf{C} \in \mathbb{R}^{1\times 2}, D \in \mathbb{R}, \mathbf{o}_0 = [0,0]^\top$, and $\sigma_k = \{1,1,1,...\}$ is the discrete-time unit step function. The combination of $H(s)$ driven by the Dirac impulse with $\ell\{\delta(t)\}$ could also have been considered as a generative model. Based on the state-space model above, modeling p sinusoidal reference inputs $\mathbf{r}_k \in \Omega_{R_m} \subset \mathbb{R}^p$ requires $2p$ states. Generally speaking, the generative model of the reference input must obey the Markov property.

Consider next that the extended state-space model that consists of (1), (2), and the state-space generative model of the reference input signal is, in the most general form [12,46]:

$$\mathbf{x}_{k+1}^{E}=\begin{bmatrix}\mathbf{x}_{k+1}\\ \mathbf{x}_{k+1}^{m}\\ \mathbf{r}_{k+1}\end{bmatrix}=\begin{bmatrix}\mathbf{f}(\mathbf{x}_k,\mathbf{u}_k)\\ \mathbf{f}^{m}(\mathbf{x}_k^{m},\mathbf{r}_k)\\ \mathbf{h}^{m}(\mathbf{r}_k)\end{bmatrix}=\mathbf{E}(\mathbf{x}_k^{E},\mathbf{u}_k),\mathbf{x}_k^{E}\in\Omega_{X^E}, \tag{4}$$

where $\mathbf{x}_k^E$ is called the extended state vector. Note that the extended state-space fulfils the Markov property. The ORM tracking control problem is formulated in an optimal control framework. Let the infinite horizon cost function (c.f.) to be minimized starting with $\mathbf{x}_0$ be [6,12,46]

$$J_{MR}^{\infty}(\mathbf{x}_0^E,\boldsymbol{\theta})=\sum_{k=0}^{\infty}\gamma^k\|\mathbf{y}_k^m(\mathbf{x}_k^E)-\mathbf{y}_k(\mathbf{x}_k^E,\boldsymbol{\theta})\|_2^2=\sum_{k=0}^{\infty}\gamma^k\|\boldsymbol{\epsilon}_k(\mathbf{x}_k^E,\boldsymbol{\theta})\|_2^2. \tag{5}$$

In (5), the discount factor $0<\gamma\leq 1$ sets the controller's horizon, $\gamma<1$ is usually used in practice to guarantee learning convergence to optimal control. $\|\mathbf{x}\|_2=\sqrt{\mathbf{x}^\top\mathbf{x}}$ is the Euclidean norm of the column vector $\mathbf{x}$. $v_{MR}=\|\mathbf{y}_k^m(\mathbf{x}_k^E)-\mathbf{y}_k(\mathbf{x}_k^E)\|_2^2$ is the stage cost where measurable $\mathbf{y}_k$ depends via unknown $\mathbf{g}$ in (1) on $\mathbf{x}_k$ and v_{MR} penalizes the deviation of $\mathbf{y}_k$ from the ORM's output $\mathbf{y}_k^m$. In ORM tracking, the stage cost does not penalize the control effort with some positive definite function $W(\mathbf{u}_k)>0$ since the ORM tracking instills an inertia on the CS that indirectly acts as a regularizer on the control effort. Secondly, if the reference inputs $\mathbf{r}_k$ do not set to zero, the ORM's outputs also do not. For most processes, the corresponding constant steady-state control will be non-zero, hence making $J_{MR}^{\infty}(\boldsymbol{\theta})$ infinite when $\gamma=1$ [12,46].

Herein, $\boldsymbol{\theta}\in\mathbb{R}^{n_\theta}$ parameterizes a nonlinear state-feedback admissible controller [6] defined as $\mathbf{u}_k\triangleq\mathbf{C}(\mathbf{x}_k^E,\boldsymbol{\theta})$, which used in (4) shows that all CS's trajectories depend on $\boldsymbol{\theta}$. Any stabilizing controller sequence (or controller) rendering a finite c.f. is called admissible. A finite J_{MR}^{∞} holds if $\boldsymbol{\epsilon}_k$ is a square-summable sequence, ensured by an asymptotically stabilizing controller if $\gamma=1$ or by a stabilizing controller if $\gamma<1$. $J_{MR}^{\infty}(\boldsymbol{\theta})$ in (5) is the value function of using the controller $\mathbf{C}(\boldsymbol{\theta})$. Let the optimal controller $\mathbf{u}_k^*=\mathbf{C}(\mathbf{x}_k^E,\boldsymbol{\theta}^*)$ that minimizes (5) be [12,46]

$$\boldsymbol{\theta}^*=arg\min_{\boldsymbol{\theta}}J_{MR}^{\infty}(\mathbf{x}_0^E,\boldsymbol{\theta}). \tag{6}$$

Tracking a nonlinear ORM can also be used, however, tracking a linear one renders highly desirable indirect feedback linearization of the CS, where a linear CS's behavior generalizes well in wide operating ranges [1]. Then the ORM tracking control problem of this work should make $v_{MR}\approx 0$ when $\mathbf{r}_k$ drives both the CS and the ORM.

Under classical control rules, following *Comment 1*, the process time delay and non-minimum-phase (NMP) character should be accounted for in $\mathbf{M}(z)$. However, the NMP zeroes make $\mathbf{M}(z)$ non-invertible in addition to requiring their identification, thus placing a burden on the subsequent VRFT IO control design [47]. This motivates the MP assumption on the process.

Depending on the learning context, the user may select a piece-wise constant generative model for the reference input signal such as $\mathbf{r}_{k+1}=\mathbf{r}_k$, or a ramp-like model, a sine-like model, etc. In all cases, the states of the generative model are known, measurable and need to be introduced in the extended state vector, to fulfill the Markov property of the extended state-space model. In many practical applications, for the ORM tracking problem, the CS's outputs are required to track the ORM's outputs when both the ORM and the CS are driven by the piece-wise constant reference input signal expressed by a generative model of the form $\mathbf{r}_{k+1}=\mathbf{r}_k$. This generative model will be used subsequently in this paper for learning ORM tracking controllers. Obviously, the learnt solution will depend on the proposed reference input generative model, while changing this model requires re-learning.

3. Solution to the ORM Tracking Problem

For unknown extended process dynamics (4), minimization of (5) can be tackled using an iterative model-free approximate Value Iteration (IMF-AVI). A c.f. that extends $J_{MR}^{\infty}(\mathbf{x}_k^E)$ called the Q-function (or action-value function) is first defined for each state-action pair. Let the Q-function of acting as $\mathbf{u}_k$ in state $\mathbf{x}_k^E$ and then following the control (policy) $\mathbf{u}_k = \mathbf{C}(\mathbf{x}_k^E)$ be [12,46]

$$Q^{\mathbf{C}}(\mathbf{x}_k^E, \mathbf{u}_k) = v(\mathbf{x}_k^E, \mathbf{u}_k) + \gamma Q^{\mathbf{C}}(\mathbf{x}_{k+1}^E, \mathbf{C}(\mathbf{x}_{k+1}^E)). \tag{7}$$

The optimal Q-function $Q^*(\mathbf{x}_k^E, \mathbf{u}_k)$ corresponding to the optimal controller obeys Bellman's optimality equation [12,46]

$$Q^*(\mathbf{x}_k^E, \mathbf{u}_k) = \min_{\mathbf{C}(.)}\{v(\mathbf{x}_k^E, \mathbf{u}_k) + \gamma Q^{\mathbf{C}}(\mathbf{x}_{k+1}^E, \mathbf{C}(\mathbf{x}_{k+1}^E))\}, \tag{8}$$

where the optimal controller and Q-functions are [12,46]

$$\mathbf{u}_k^* = \mathbf{C}^*(\mathbf{x}_k^E) = arg\min_{\mathbf{C}} Q^{\mathbf{C}}(\mathbf{x}_k^E, \mathbf{u}_k), Q^*(\mathbf{x}_k^E, \mathbf{u}_k) = \min_{\mathbf{C}(.)} Q^{C}(\mathbf{x}_k^E, \mathbf{u}_k). \tag{9}$$

Then, for $J_{MR}^{\infty,*} = \min_{\mathbf{u}} J_{MR}^{\infty}(\mathbf{x}_k^E, \mathbf{u}_k)$ it follows that $J_{MR}^{\infty,*} = Q^*(\mathbf{x}_k^E, \mathbf{u}_k^* = \mathbf{C}^*(\mathbf{x}_k^E))$. Implying that finding Q^* is equivalent to finding the optimal c.f. $J_{MR}^{\infty,*}$.

The optimal Q-function and optimal controller can be found using either Policy Iteration (PoIt) or Value Iteration (VI) strategies. For continuous state-action spaces, IMF-AVI is one possible solution, using different linear and/or nonlinear parameterizations for the Q-function and/or for the controller. NNs are most widely used as nonlinearly parameterized function approximators. As it is well-known, VI alternates two steps: the Q-function estimate update step and the controller improvement step. Several Q-function parameterizations allow for explicit analytic calculation of the improved controller as the following optimization problem (© 2019 IEEE [12])

$$\tilde{\mathbf{C}}(\mathbf{x}_k^E, \boldsymbol{\pi}) = arg\min_{\mathbf{C}} Q^{\mathbf{C}}(\mathbf{x}_k^E, \mathbf{u}_k, \boldsymbol{\pi}), \tag{10}$$

by directly minimizing $Q^{\mathbf{C}}(\mathbf{x}_k^E, \mathbf{u}_k, \boldsymbol{\pi})$ w.r.t. $\mathbf{u}_k$, where the parameterization $\boldsymbol{\pi}$ has been moved from the controller into the Q-function. (10) is the controller improvement step specific to both the PoIt and VI algorithms. In these special cases, it is possible to eliminate the controller approximator and use only one for the Q-function Q. Then, given a dataset D of transition samples, $D = \{(\mathbf{x}_k^E, \mathbf{u}_k, \mathbf{x}_{k+1}^E)\}, k = \overline{1,N}$ the IMF-AVI amounts to solving the following optimization problem (OP) at every iteration j (© 2019 IEEE [12])

$$\boldsymbol{\pi}_{j+1} = arg\min_{\boldsymbol{\pi}} \sum_{k=1}^{N} (Q(\mathbf{x}_k^E, \mathbf{u}_k, \boldsymbol{\pi}) - v(\mathbf{x}_k^E, \mathbf{u}_k) - \gamma Q(\mathbf{x}_{k+1}^E, \tilde{\mathbf{C}}(\mathbf{x}_{k+1}^E, \boldsymbol{\pi}_j), \boldsymbol{\pi}_j))^2, \tag{11}$$

which is a Bellman residual minimization problem where the (usually separate) controller improvement step is now embedded inside the OP (11). More explicitly, for a linear parameterization $Q(\mathbf{x}_k^E, \mathbf{u}_k, \boldsymbol{\pi}) = \boldsymbol{\Phi}^\top(\mathbf{x}_k^E, \mathbf{u}_k)\boldsymbol{\pi}$ using a set of n_Φ basis functions of the form $\boldsymbol{\Phi}^\top(\mathbf{x}_k^E, \mathbf{u}_k) = [\Phi_1(\mathbf{x}_k^E, \mathbf{u}_k), ..., \Phi_{n_\Phi}(\mathbf{x}_k^E, \mathbf{u}_k)]$, the least squares solution to (11) is equivalent to solving the following over-determined linear system of equations w.r.t. $\boldsymbol{\pi}_{j+1}$ in the least-squares sense (© 2019 IEEE [12]):

$$\begin{bmatrix} \boldsymbol{\Phi}^\top(\mathbf{x}_1^E, \mathbf{u}_1) \\ \dots \\ \boldsymbol{\Phi}^\top(\mathbf{x}_N^E, \mathbf{u}_N) \end{bmatrix} \boldsymbol{\pi}_{j+1} = \begin{bmatrix} v(\mathbf{x}_1^E, \mathbf{u}_1) + \gamma\boldsymbol{\Phi}^\top(\mathbf{x}_2^E, \tilde{\mathbf{C}}(\mathbf{x}_2^E, \boldsymbol{\pi}_j))\boldsymbol{\pi}_j \\ \dots \\ v(\mathbf{x}_N^E, \mathbf{u}_N) + \gamma\boldsymbol{\Phi}^\top(\mathbf{x}_{N+1}^E, \tilde{\mathbf{C}}(\mathbf{x}_{N+1}^E, \boldsymbol{\pi}_j))\boldsymbol{\pi}_j \end{bmatrix}. \tag{12}$$

Concluding, starting with an initial parameterization $\boldsymbol{\pi}_0$, the IMF-AVI approach with linearly parameterized Q-function that allows explicit controller improvement calculation as in (10), embeds both VI steps into solving (12). Linearly parameterized IMF-AVI (LP-IMF-AVI) will be validated in the case study and compared to nonlinearly parameterized IMF-AVI (NP-IMF-AVI). Convergence of the generally formulated IMF-AVI is next analysed under approximation errors (© 2019 IEEE [12]).

IMF-AVI Convergence Analysis with Approximation Errors for ORM Tracking

The proposed iterative model-free VI-based Q-learning Algorithm 1 consists of the next steps (© 2019 IEEE [12]).

Algorithm 1 VI-based Q-learning.

S1: Initialize controller $\mathbf{C}_0$ and the Q-function value to $Q_0(\mathbf{x}_k^E, \mathbf{u}_k) = 0$, initialize iteration index $j = 1$
S2: Use one step backup equation for the Q-function as in (13)
S3: Improve the controller using the Equation (14)
S4: Set $j = j + 1$ and repeat steps S2, S3, until convergence

To be detailed as follows:

S1. Select an initial (not necessarily admissible) controller $\mathbf{C}_0$ and an initialization value $Q_0(\mathbf{x}_k^E, \mathbf{u}_k) = 0$ of the Q-function. Initialize iteration $j = 1$.

S2. Use one step backup equation for the Q-function

$$\begin{aligned} Q_j(\mathbf{x}_k^E, \mathbf{u}_k) &= v(\mathbf{x}_k^E, \mathbf{u}_k) + \gamma Q_{j-1}(\mathbf{x}_{k+1}^E, \mathbf{C}_{j-1}(\mathbf{x}_{k+1}^E)) \\ &= \min_{\mathbf{u}}\{v(\mathbf{x}_k^E, \mathbf{u}_k) + \gamma Q_{j-1}(\mathbf{x}_{k+1}^E, \mathbf{u})\} \end{aligned} \tag{13}$$

S3. Improve the controller using the equation

$$\mathbf{C}_j(\mathbf{x}_k^E) = arg \min_{\mathbf{u}} Q_j(\mathbf{x}_k^E, \mathbf{u}). \tag{14}$$

S4. Set $j = j + 1$ and repeat steps S2, S3, until convergence.

Lemma 1. *(© 2019 IEEE [12]) For an arbitrary sequence of controllers $\{\boldsymbol{\kappa}_j\}$ define the VI-update for extended c.f. ξ_j as [48]*

$$\xi_{j+1}(\mathbf{x}_k^E, \mathbf{u}_k) = v(\mathbf{x}_k^E, \mathbf{u}_k) + \gamma \xi_j(\mathbf{x}_{k+1}^E, \boldsymbol{\kappa}_j(\mathbf{x}_{k+1}^E)). \tag{15}$$

If $Q_0(\mathbf{x}_k^E, \mathbf{u}_k) = \xi_0(\mathbf{x}_k^E, \mathbf{u}_k) = 0$, then $Q_j(\mathbf{x}_k^E, \mathbf{u}_k) \le \xi_j(\mathbf{x}_k^E, \mathbf{u}_k)$.

Proof. It is valid that

$$\begin{aligned} Q_1(\mathbf{x}_k^E, \mathbf{u}_k) &= v(\mathbf{x}_k^E, \mathbf{u}_k) + \gamma \overbrace{Q_0(\mathbf{x}_{k+1}^E, \mathbf{C}_0(\mathbf{x}_{k+1}^E))}^{0} = \\ &= v(\mathbf{x}_k^E, \mathbf{u}_k) + \gamma \overbrace{\xi_0(\mathbf{x}_{k+1}^E, \boldsymbol{\kappa}_0(\mathbf{x}_{k+1}^E))}^{0} = \xi_1(\mathbf{x}_k^E, \mathbf{u}_k). \end{aligned} \tag{16}$$

Meaning that $Q_1(\mathbf{x}_k^E, \mathbf{u}_k) \le \xi_1(\mathbf{x}_k^E, \mathbf{u}_k)$. Assume by induction that $Q_{j-1}(\mathbf{x}_k^E, \mathbf{u}_k) \le \xi_{j-1}(\mathbf{x}_k^E, \mathbf{u}_k)$. Then

$$\begin{aligned} Q_j(\mathbf{x}_k^E, \mathbf{u}_k) &= v(\mathbf{x}_k^E, \mathbf{u}_k) + \gamma Q_{j-1}(\mathbf{x}_{k+1}^E, \mathbf{C}_{j-1}(\mathbf{x}_{k+1}^E)) \le \\ &\le v(\mathbf{x}_k^E, \mathbf{u}_k) + \gamma Q_{j-1}(\mathbf{x}_{k+1}^E, \boldsymbol{\kappa}_{j-1}(\mathbf{x}_{k+1}^E)) \le \\ &\le v(\mathbf{x}_k^E, \mathbf{u}_k) + \gamma \xi_{j-1}(\mathbf{x}_{k+1}^E, \boldsymbol{\kappa}_{j-1}(\mathbf{x}_{k+1}^E)) = \xi_j(\mathbf{x}_k^E, \mathbf{u}_k), \end{aligned} \tag{17}$$

which completes the proof. Here, it was used that $\mathbf{C}_{j-1}(\mathbf{x}_k^E)$ is the optimal controller for $Q_{j-1}(\mathbf{x}_k^E, \mathbf{u}_k)$ per (14), then, for any other controller $\mathbf{C}(\mathbf{x}_k^E)$ (in particular it can also be $\boldsymbol{\kappa}_{j-1}(\mathbf{x}_k^E)$) it follows that

$$Q_{j-1}(\mathbf{x}_{k+1}^E, \mathbf{C}_{j-1}(\mathbf{x}_{k+1}^E)) \le Q_{j-1}(\mathbf{x}_{k+1}^E, \mathbf{C}(\mathbf{x}_{k+1}^E)). \tag{18}$$

□

Lemma 2. *(© 2019 IEEE [12]) For the sequence $\{Q_j\}$ from (13), under controllability assumption A1, it is valid that:*

(1) $0 \le Q_j(x_k^E, u_k) \le B(x_k^E, u_k)$ *with* $B(x_k^E, u_k)$ *an upper bound.*
(2) If there exists a solution $Q^*(x_k^E, u_k)$ *to (8), then* $0 \le Q_j(x_k^E, u_k) \le Q^*(x_k^E, u_k) \le B(x_k^E, u_k)$.

Proof. For any fixed admissible controller $\boldsymbol{\eta}(\mathbf{x}_k^E)$, $Q^{\eta}(\mathbf{x}_k^E, \mathbf{u}_k) = v(\mathbf{x}_k^E, \mathbf{u}_k) + \gamma Q^{\eta}(\mathbf{x}_{k+1}^E, \boldsymbol{\eta}(\mathbf{x}_{k+1}^E))$ is the Bellman equation. Update (13) renders

$$
\begin{aligned}
Q_j(\mathbf{x}_k^E, \mathbf{u}_k) &= v(\mathbf{x}_k^E, \mathbf{u}_k) + \gamma Q_{j-1}(\mathbf{x}_{k+1}^E, \mathbf{C}_{j-1}(\mathbf{x}_{k+1}^E)) \overset{(18)}{\le} \\
&\overset{(18)}{\le} v(\mathbf{x}_k^E, \mathbf{u}_k) + \gamma Q_{j-1}(\mathbf{x}_{k+1}^E, \boldsymbol{\eta}(\mathbf{x}_{k+1}^E)) \\
Q_{j-1}(\mathbf{x}_{k+1}^E, \boldsymbol{\eta}(\mathbf{x}_{k+1}^E)) &= v(\mathbf{x}_{k+1}^E, \boldsymbol{\eta}(\mathbf{x}_{k+1}^E)) + \gamma Q_{j-2}(\mathbf{x}_{k+2}^E, \mathbf{C}_{j-2}(\mathbf{x}_{k+2}^E)) \overset{(18)}{\le} \\
&\overset{(18)}{\le} v(\mathbf{x}_{k+1}^E, \boldsymbol{\eta}(\mathbf{x}_{k+1}^E)) + \gamma Q_{j-2}(\mathbf{x}_{k+2}^E, \boldsymbol{\eta}(\mathbf{x}_{k+2}^E)) \\
&\dots \\
Q_1(\mathbf{x}_{k+j-1}^E, \boldsymbol{\eta}(\mathbf{x}_{k+j-1}^E)) &= v(\mathbf{x}_{k+j-1}^E, \boldsymbol{\eta}(\mathbf{x}_{k+j-1}^E)) + \gamma \overbrace{Q_0(\mathbf{x}_{k+j}^E, \mathbf{C}_0(\mathbf{x}_{k+j}^E))}^{0}
\end{aligned}
\tag{19}
$$

Replacing from the last inequality towards the first it follows that

$$
\begin{aligned}
Q_j(\mathbf{x}_k^E, \mathbf{u}_k) &\le v(\mathbf{x}_k^E, \mathbf{u}_k) + \gamma v(\mathbf{x}_{k+1}^E, \boldsymbol{\eta}(\mathbf{x}_{k+1}^E)) + \dots + \gamma^{j-1}(\mathbf{x}_{k+j-1}^E, \boldsymbol{\eta}(\mathbf{x}_{k+j-1}^E)) \\
&\le v(\mathbf{x}_k^E, \mathbf{u}_k) + \sum_{j=1}^{\infty} \gamma^j v(\mathbf{x}_{k+j}^E, \boldsymbol{\eta}(\mathbf{x}_{k+j}^E)) = Q^{\eta}(\mathbf{x}_k^E, \mathbf{u}_k),
\end{aligned}
\tag{20}
$$

then, setting $Q^{\eta}(\mathbf{x}_k^E, \mathbf{u}_k) = B(\mathbf{x}_k^E, \mathbf{u}_k)$ proves the first part of Lemma 2.

Among all admissible controllers, the optimal one renders the Q-function with the lowest value therefore $Q^*(\mathbf{x}_k^E, \mathbf{u}_k) \le Q^{\eta}(\mathbf{x}_k^E, \mathbf{u}_k) = B(\mathbf{x}_k^E, \mathbf{u}_k)$. If $\boldsymbol{\eta}(\mathbf{x}_k^E) = \mathbf{C}^*(\mathbf{x}_k^E)$ is the optimal controller, it follows that $Q_j(\mathbf{x}_k^E, \mathbf{u}_k) \le Q^*(\mathbf{x}_k^E, \mathbf{u}_k)$. Then the second part of Lemma 2 follows as $0 \le Q_j(\mathbf{x}_k^E, \mathbf{u}_k) \le Q^*(\mathbf{x}_k^E, \mathbf{u}_k) \le B(\mathbf{x}_k^E, \mathbf{u}_k)$. □

Theorem 1. *(© 2019 IEEE [12]) For the extended process (4) under A1, A2, with c.f. (5), with the sequences* $\{C_j\}$ *and* $\{Q_j(x_k^E, u_k)\}$ *generated by the Q-learning Algorithm 1, it is true that:*
(1) $\{Q_j(x_k^E, u_k)\}$ *is a non-decreasing sequence for which* $Q_{j+1}(x_k^E, u_k) \ge Q_j(x_k^E, u_k)$ *holds,* $\forall j, \forall (x_k^E, u_k)$ *and*
(2) $\lim_{j\to\infty} C_j(x_k^E) = C^*(x_k^E)$ *and* $\lim_{j\to\infty} Q_j(x_k^E, u_k) = Q^*(x_k^E, u_k)$.

Proof. Let $Q_0(\mathbf{x}_k^E, \mathbf{u}_k) = \xi_0(\mathbf{x}_k^E, \mathbf{u}_k) = 0$ and assume the update

$$
\xi_j(\mathbf{x}_k^E, \mathbf{u}_k) = v(\mathbf{x}_k^E, \mathbf{u}_k) + \gamma \xi_{j-1}(\mathbf{x}_{k+1}^E, \mathbf{C}_j(\mathbf{x}_{k+1}^E)). \tag{21}
$$

By induction it is shown that $Q_1(\mathbf{x}_k^E, \mathbf{u}_k) \ge \xi_0(\mathbf{x}_k^E, \mathbf{u}_k)$ since

$$
Q_1(\mathbf{x}_k^E, \mathbf{u}_k) = v(\mathbf{x}_k^E, \mathbf{u}_k) + \gamma Q_0(\mathbf{x}_{k+1}^E, \mathbf{C}_0(\mathbf{x}_{K+1}^E)) = v(\mathbf{x}_k^E, \mathbf{u}_k) + \gamma \cdot 0 \ge 0 = \xi_0(\mathbf{x}_k^E, \mathbf{u}_k). \tag{22}
$$

Assume next that $Q_j(\mathbf{x}_k^E, \mathbf{u}_k) \ge \xi_{j-1}(\mathbf{x}_k^E, \mathbf{u}_k)$ and show that

$$
\begin{aligned}
&Q_{j+1}(\mathbf{x}_k^E, \mathbf{u}_k) - \xi_j(\mathbf{x}_k^E, \mathbf{u}_k) = v(\mathbf{x}_k^E, \mathbf{u}_k) + \gamma Q_j(\mathbf{x}_{k+1}^E, \mathbf{C}_j(\mathbf{x}_{k+1}^E)) - v(\mathbf{x}_k^E, \mathbf{u}_k) - \\
&\gamma \xi_{j-1}(\mathbf{x}_{k+1}^E, \mathbf{C}_j(\mathbf{x}_{k+1}^E)) = \gamma [Q_j(\mathbf{x}_{k+1}^E, \mathbf{C}_j(\mathbf{x}_{k+1}^E)) - \xi_{j-1}(\mathbf{x}_{k+1}^E, \mathbf{C}_j(\mathbf{x}_{k+1}^E))] \ge 0.
\end{aligned}
\tag{23}
$$

The expression above leads to $Q_{j+1}(\mathbf{x}_k^E, \mathbf{u}_k) \geq \xi_j(\mathbf{x}_k^E, \mathbf{u}_k)$. Since by Lemma 1 one has that $Q_j(\mathbf{x}_k^E, \mathbf{u}_k) \geq \xi_j(\mathbf{x}_k^E, \mathbf{u}_k)$ it follows that $Q_{j+1}(\mathbf{x}_k^E, \mathbf{u}_k) \geq Q_j(\mathbf{x}_k^E, \mathbf{u}_k)$, proving first part of Theorem 1.

Any non-decreasing upper bounded sequence must have a limit, thus $\lim_{j\to\infty} \mathbf{C}_j = \mathbf{C}_\infty$ and $\lim_{j\to\infty} Q_j(\mathbf{x}_k^E, \mathbf{u}_k) = Q_\infty(\mathbf{x}_k^E, \mathbf{u}_k)$ with $\mathbf{C}_\infty$ an admissible controller. For any admissible controller $\eta(\mathbf{x}_k^E) = \mathbf{C}_\infty(\mathbf{x}_k^E)$ that is non-optimal if follows from (20) that $Q^*(\mathbf{x}_k^E, \mathbf{u}_k) \leq Q_\infty(\mathbf{x}_k^E, \mathbf{u}_k)$. Still, part 2 of Lemma 2 states that $Q_j(\mathbf{x}_k^E, \mathbf{u}_k) \leq Q^*(\mathbf{x}_k^E, \mathbf{u}_k)$ implying $Q_\infty(\mathbf{x}_k^E, \mathbf{u}_k) \leq Q^*(\mathbf{x}_k^E, \mathbf{u}_k)$. Then from $Q_\infty(\mathbf{x}_k^E, \mathbf{u}_k) \leq Q^*(\mathbf{x}_k^E, \mathbf{u}_k) \leq Q_\infty(\mathbf{x}_k^E, \mathbf{u}_k)$ it must hold true that $Q_\infty(\mathbf{x}_k^E, \mathbf{u}_k) = Q^*(\mathbf{x}_k^E, \mathbf{u}_k)$ and $\mathbf{C}_\infty(\mathbf{x}_k^E) = \mathbf{C}^*(\mathbf{x}_k^E)$ which proves the second part of Theorem 1. □

Comment 2. (© 2019 IEEE [12]) (13) is practically solved in the sense of the OP (11) (either as a linear or nonlinear regression) using a batch (dataset) of transition samples collected from the process using any controller, that is in *off-policy* mode. While the controller improvement step (14) can be solved either as a regression or explicitly analytically when the expression of $Q_j(\mathbf{x}_k^E, \mathbf{u}_k)$ allows it. Moreover, (13) and (14) can be solved batch-wise in either online or offline mode. When the batch of transition samples is updated with one sample at a time, the VI-scheme becomes adaptive.

Comment 3. (© 2019 IEEE [12]) Theorem 1 proves the VI-based learning convergence of the sequence of Q-functions $\lim_{j\to\infty} Q_j(\mathbf{x}_k^E, \mathbf{u}_k) = Q^*(\mathbf{x}_k^E, \mathbf{u}_k)$ assuming that the true Q-function parameterization is used. In practice, this is rarely possible, such as, e.g., in the case of LTI systems. For general nonlinear processes of type (1), different function approximators are employed for the Q-function, most commonly using NNs. Then the convergence of the VI Q-learning scheme is to a suboptimal controller and to a suboptimal Q-function, owing to the approximation errors. A generic convergence proof of the learning scheme under approximation errors is next shown, accounting for general Q-function parameterizations [49].

Let the IMF-AVI Algorithm 2 consist of the steps (© 2019 IEEE [12]).

Algorithm 2 IMF-AVI.

S1: Initialize controller $\tilde{\mathbf{C}}_0$ and Q-function value $\tilde{Q}_0(\mathbf{x}_k^E, \mathbf{u}_k) = 0, \forall(\mathbf{x}_k^E, \mathbf{u}_k)$. Initialize iteration $j = 1$
S2: Update the approximate Q-function using Equation (24)
S3: Improve the approximate controller using Equation (25)
S4: Set $j = j + 1$ and repeat steps S2, S3, until convergence

To be detailed as follows:

S1. Select an initial (not necessarily admissible) controller $\tilde{\mathbf{C}}_0$ and an initialization value $\tilde{Q}_0(\mathbf{x}_k^E, \mathbf{u}_k) = 0, \forall(\mathbf{x}_k^E, \mathbf{u}_k)$ of the Q-function. Initialize iteration $j = 1$.

S2. Use the update equation for the approximate Q-function

$$\begin{aligned}\tilde{Q}_j(\mathbf{x}_k^E, \mathbf{u}_k) &= v(\mathbf{x}_k^E, \mathbf{u}_k) + \gamma \tilde{Q}_{j-1}(\mathbf{x}_{k+1}^E, \tilde{\mathbf{C}}_{j-1}(\mathbf{x}_{k+1}^E)) + \delta_j(\mathbf{x}_k^E, \mathbf{u}_k) \\ &= \min_{\mathbf{u}}\{v(\mathbf{x}_k^E, \mathbf{u}_k) + \gamma \tilde{Q}_{j-1}(\mathbf{x}_{k+1}^E, \mathbf{u})\} + \delta_j\end{aligned} \tag{24}$$

S3. Improve the approximate controller using

$$\tilde{\mathbf{C}}_j(\mathbf{x}_k^E) = arg \min_{\mathbf{u}} \tilde{Q}_j(\mathbf{x}_k^E, \mathbf{u}) \tag{25}$$

S4. Set $j = j + 1$ and repeat steps S2, S3, until convergence.

Comment 4. (© 2019 IEEE [12]) In Algorithm 2, the sequences $\{\tilde{\mathbf{C}}_j(\mathbf{x}_k^E)\}$ and $\{\tilde{Q}_j(\mathbf{x}_k^E, \mathbf{u})\}$ are approximations of the true sequences $\{\mathbf{C}_j(\mathbf{x}_k^E)\}$ and $\{Q_j(\mathbf{x}_k^E, \mathbf{u}_k)\}$. Since the true Q-function and controller parameterizations are not generally known, (24) must be solved in the sense of the OP (11) with respect to the unknown $\tilde{Q}_j$, in order to minimize the residuals δ_j at each iteration. If the true parameterizations of the Q-function and of the controller were known, then $\delta_j = 0$ and the IMF-AVI updates (24), (25) coincide with (13), (14), respectively. Next, let the following assumption hold. A4. ([12]) There exist two positive scalar constants $\underline{\psi}, \overline{\psi}$ such that $0 < \underline{\psi} \leq 1 \leq \overline{\psi} < \infty$, ensuring

$$\min_{\mathbf{u}}\{\underline{\psi}v(\mathbf{x}_k^E, \mathbf{u}_k) + \gamma\tilde{Q}_{j-1}(\mathbf{x}_{k+1}^E, \mathbf{u})\} \leq \tilde{Q}_j(\mathbf{x}_k^E, \mathbf{u}_k) \leq \min_{\mathbf{u}}\{\overline{\psi}v(\mathbf{x}_k^E, \mathbf{u}_k) + \gamma\tilde{Q}_{j-1}(\mathbf{x}_{k+1}^E, \mathbf{u})\}. \quad (26)$$

Comment 5. (© 2019 IEEE [12]) Inequalities from (26) account for nonzero positive or negative residuals δ_j, i.e., for the approximation errors in the Q-function, since $\tilde{Q}_j(\mathbf{x}_k^E, \mathbf{u}_k)$ can over- or under-estimate $\min_{\mathbf{u}}\{v(\mathbf{x}_k^E, \mathbf{u}_k) + \gamma\tilde{Q}_{j-1}(\mathbf{x}_{k+1}^E, \mathbf{u})\}$ in (24). $\underline{\psi}, \overline{\psi}$ can span large intervals ($\underline{\psi}$ close to 0 and $\overline{\psi}$ very large). The hope is that, if $\underline{\psi}, \overline{\psi}$ are close to 1—meaning low approximation errors—then the entire IMF-AVI process preserves $\delta_j \approx 0$. In practice, this amounts to using high performance approximators. For example, with NNs, adding more layers and more neurons, enhances the approximation capability and theoretically reduces the residuals in (24).

Theorem 2. *Let the sequences* $\{\tilde{\mathbf{C}}_j(\mathbf{x}_k^E)\}$ *and* $\{\tilde{Q}_j(\mathbf{x}_k^E, \mathbf{u}_k)\}$ *evolve as in (24), (25), the sequences* $\{\mathbf{C}_j(\mathbf{x}_k^E)\}$ *and* $\{Q_j(\mathbf{x}_k^E, \mathbf{u}_k)\}$ *evolve as in (13), (14). Initialize* $\tilde{Q}_0(\mathbf{x}_k^E, \mathbf{u}_k) = Q_0(\mathbf{x}_k^E, \mathbf{u}_k) = 0, \forall(\mathbf{x}_k^E, \mathbf{u}_k)$ *and let A3 hold. Then*

$$\underline{\psi}Q_j(\mathbf{x}_k^E, \mathbf{u}_k) \leq \tilde{Q}_j(\mathbf{x}_k^E, \mathbf{u}_k) \leq \overline{\psi}Q_j(\mathbf{x}_k^E, \mathbf{u}_k) \quad (27)$$

Proof. (© 2019 IEEE [12]) First, the development proceeds by induction for the left inequality. For $j = 0$ it is clear that $\underline{\psi}Q_0(\mathbf{x}_k^E, \mathbf{u}_k) \leq \tilde{Q}_0(\mathbf{x}_k^E, \mathbf{u}_k)$. For $j = 1$, (13) produces $Q_1(\mathbf{x}_k^E, \mathbf{u}_k) = v(\mathbf{x}_k^E, \mathbf{u}_k)$ and left-hand side of (26) reads $\min_{\mathbf{u}}\{\underline{\psi}v(\mathbf{x}_k^E, \mathbf{u}_k) + 0\} \leq \tilde{Q}_1(\mathbf{x}_k^E, \mathbf{u}_k)$. Then $\underline{\psi}Q_1(\mathbf{x}_k^E, \mathbf{u}_k) \leq \tilde{Q}_1(\mathbf{x}_k^E, \mathbf{u}_k)$. Next assume that

$$\underline{\psi}Q_j(\mathbf{x}_k^E, \mathbf{u}_k) \leq \tilde{Q}_j(\mathbf{x}_k^E, \mathbf{u}_k) \quad (28)$$

holds at iteration j. Based on (28) used in (26), it is valid that

$$\min_{\mathbf{u}}\{\underline{\psi}v(\mathbf{x}_k^E, \mathbf{u}_k) + \gamma\underline{\psi}Q_j(\mathbf{x}_{k+1}^E, \mathbf{u})\} \leq \min_{\mathbf{u}}\{\underline{\psi}v(\mathbf{x}_k^E, \mathbf{u}_k) + \gamma\tilde{Q}_j(\mathbf{x}_{k+1}^E, \mathbf{u})\} \leq \tilde{Q}_{j+1}(\mathbf{x}_k^E, \mathbf{u}_k). \quad (29)$$

Notice from (29) that

$$\min_{u}\{\underline{\psi}v(\mathbf{x}_k^E, \mathbf{u}_k) + \gamma\underline{\psi}Q_j(\mathbf{x}_{k+1}^E, \mathbf{u})\} = \underline{\psi}\min_{u}\{v(\mathbf{x}_k^E, \mathbf{u}_k) + \gamma Q_j(\mathbf{x}_{k+1}^E, \mathbf{u})\} \overset{(13)}{=} \underline{\psi}Q_{j+1}(\mathbf{x}_k^E, \mathbf{u}_k) \quad (30)$$

From (29), (30) it follows that $\underline{\psi}Q_{j+1}(\mathbf{x}_k^E, \mathbf{u}_k) \leq \tilde{Q}_{j+1}(\mathbf{x}_k^E, \mathbf{u}_k)$ proving the left side of (27) by induction. The right side of (27) is shown similarly, proving Theorem 2. □

Comment 6. (© 2019 IEEE [12]) Theorem 2 shows that the trajectory of $\{\tilde{Q}_j(\mathbf{x}_k^E, \mathbf{u}_k)\}$ closely follows that of $\{Q_j(\mathbf{x}_k^E, \mathbf{u}_k)\}$ in a bandwidth set by $\underline{\psi}, \overline{\psi}$. It does not ensure that $\{\tilde{Q}_j(\mathbf{x}_k^E, \mathbf{u}_k)\}$ converges to a steady-state value, but in the worst case, it oscillates around $Q^*(\mathbf{x}_k^E, \mathbf{u}_k) = lim_{j\to\infty}Q_j(\mathbf{x}_k^E, \mathbf{u}_k)$ in a band that can be made arbitrarily small by using powerful approximators. By minimizing over $\mathbf{u}_k$ both sides of (27), similar conclusions result for the controller sequence $\{\tilde{\mathbf{C}}_j(\mathbf{x}_k^E)\}$ that closely follows $\{\mathbf{C}_j(\mathbf{x}_k^E)\}$.

In the following Section, the IMF-AVI is validated on two illustrative examples. The provided theoretical analysis supports and explains the robust learning performance of the nonlinearly parameterized IMF-AVI with respect to the linearly parameterized one.

4. Validation Case Studies

4.1. ORM Tracking for a Linear Process

A first introductory simple example of IMF-AVI for the ORM tracking of a first-order process motivates the more complex validation for the TITOAS process and offers insight into how the IMF-AVI solution scales up with the higher-order processes.

Let a scalar discrete-time process discretized at $T_s = 0.1s$ be $x_{k+1} = 0.8187x_k + 0.1813u_k$. The continuous-time ORM $M(s) = 1/(s+1)$ ZOH discretized at the same T_s leads to the extended process equivalent to (4), (output equations also given):

$$\begin{cases} x_{k+1} = 0.8187x_k + 0.1813u_k, \\ x^m_{k+1} = 0.9048x^m_k + 0.09516r_k, \\ r_{k+1} = r_k, \\ y_k = x_k, y^m_k = x^m_k, \end{cases} \Leftrightarrow \mathbf{x}^E_{k+1} = \mathbf{E}(\mathbf{x}^E_k, u_k) \tag{31}$$

where a piece-wise constant reference input generative model is introduced to ensure that the extended process (31) has full measurable state.

For data collection, the ORM's output y^m_k is collected along with: u_k, x_k and the reference input r_k. The measurable extended state vector is then $\mathbf{x}^E_k = [x_k, x^m_k, r_k]^\top$. A discretized version of an integral controller with t.f. $0.25/s$ at sampling period $T_s = 0.1s$ closes the loop of the control system and asymptotically stabilizes it, while calculating the control input u_k based on the feedback error $e_k = r_k - y_k$. This CS setup is used for collecting transition samples of the form $D = \{(\mathbf{x}^E_k, \mathbf{u}_k, \mathbf{x}^E_{k+1})\}$. Data is collected for 500 s, with normally distributed random reference inputs having variance $\sigma^2_r = 0.0951$, modeled as piece-wise constant steps that change their values every 20 s. Normally distributed white noise having variance $\sigma^2_u = 4.96$ is added on the command u_k at every time step to ensure a proper exploration by visiting as many combinations of states and actions as possible. Exploration has a critical role in the success of the IMF-AVI. A higher amplitude additive noise on u_k increases the chances of converging the approximate VI approach. The state transitions data collection is shown in Figure 1 for the first 1000 samples (100 s).

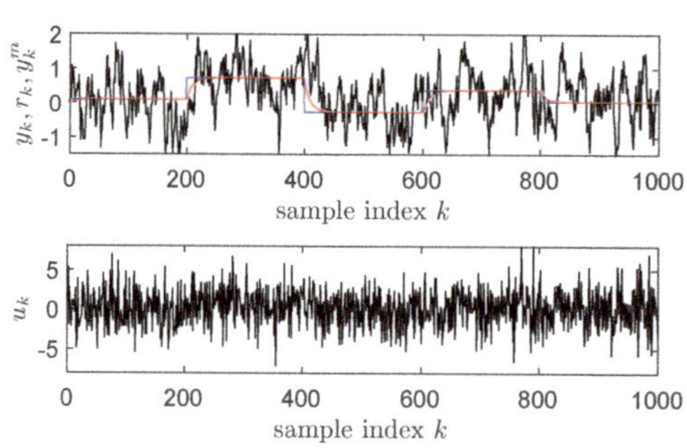

Figure 1. Closed-loop state transitions data collection for Example 1: (top) y_k (black), r_k (blue), y^m_k (red); (bottom) u_k.

Notice that a reference input modeled as a sequence of constant amplitude steps is used for exploration purposes, for which it may not be possible to write $r_{k+1} = h^m(r_k)$ as a generative model. To solve this, all transition samples that correspond to the switching times of the reference input are eliminated, therefore, $r_{k+1} = r_k$ can be considered as the piece-wise constant generative model of the reference input.

The control objective is to minimize $J^{\infty}_{MR}(\boldsymbol{\theta})$ from (5) using the stage cost $v(\mathbf{x}^E_k) = (y^m_k - y_k)^2$ (where the outputs obviously depend on the extended states as per (31)), with the discount factor

$\gamma = 0.9$. Thus the overall objective is to find the optimal state-feedback controller $\mathbf{u}_k^* = \mathbf{C}^*(\mathbf{x}_k^E)$ that makes the feedback CS match the ORM.

The Q-function is linearly parameterized as $Q(x_k^E, u_k) = \mathbf{\Phi}^\top(\mathbf{x}_k^E, u_k)\boldsymbol{\pi}$, with the quadratic basis functions vector constructed by the unique terms of the Kronecker product of all input arguments of $Q(\mathbf{x}_k^E, u_k)$ as

$$\mathbf{\Phi}^\top(\mathbf{x}_k^E, u_k) = [x_k^2, (x_k^m)^2, r_k^2, u_k^2, x_k x_k^m, x_k r_k, x_k u_k, x_k^m r_k, x_k^m u_k, r_k u_k] \tag{32}$$

with $\boldsymbol{\pi} \in \mathbb{R}^{10}$. The controller improvement step equivalent to explicitly minimizing the Q-function w.r.t. the control input u_k is $\tilde{u}_k^* = \tilde{C}^*(\mathbf{x}_k^E, \boldsymbol{\pi}_j) = -\frac{1}{2\pi_{4,j}}[\pi_{7,j}, \pi_{9,j}, \pi_{10,j}]\mathbf{x}_k^E$. This improved linear-in-the-state controller is embedded in the linear system of equations (12) that is solved for every iteration of IMF-AVI. Each iteration produces a new $\boldsymbol{\pi}_{j+1}$ that is tested on a test scenario where the uniformly random reference inputs have amplitude $r_k \in [-1; 1]$ and switch every 10 s. The ORM tracking performance is then measured by the Euclidean vector norm $\|y_k^m - y_k\|_2$ while $\|\boldsymbol{\pi}_{j+1} - \boldsymbol{\pi}_j\|_2$ serves as a stopping condition when it drops below a prescribed threshold. The practically observed convergence process is shown in Figure 2 over the first 400 iterations, with $\|\boldsymbol{\pi}_{j+1} - \boldsymbol{\pi}_j\|_2$ still decreasing after 1000 iterations. While $\|y_k^m - y_k\|_2$ is very small right from the first iterations, making the process output practically overlap with the ORM's output.

Comment 7. For LTI processes with an LQR-like c.f., an LTI ORM and an LTI generative reference input model, linear parameterizations of the extended Q-function of the form $Q(\mathbf{x}_k^E, \mathbf{u}_k) = \Phi^\top(\mathbf{x}_k^E, \mathbf{u}_k)\boldsymbol{\pi}$ is the well-known [9] form $Q(\mathbf{x}_k^E, \mathbf{u}_k) = [(\mathbf{x}_k^E)^\top, (\mathbf{u}_k)^\top]\mathbf{P}[(\mathbf{x}_k^E)^\top, (\mathbf{u}_k)^\top]^\top$ of the quadratic Q-function, with parameter $\boldsymbol{\pi} = vec(\mathbf{P})$ being the vectorized form of the symmetric positive-definite matrix $\mathbf{P}$ and the basis function vector $\Phi^\top(\mathbf{x}_k^E, \mathbf{u}_k)$ is obtained by the nonrepeatable terms of the Kronecker product of all the Q-function input arguments.

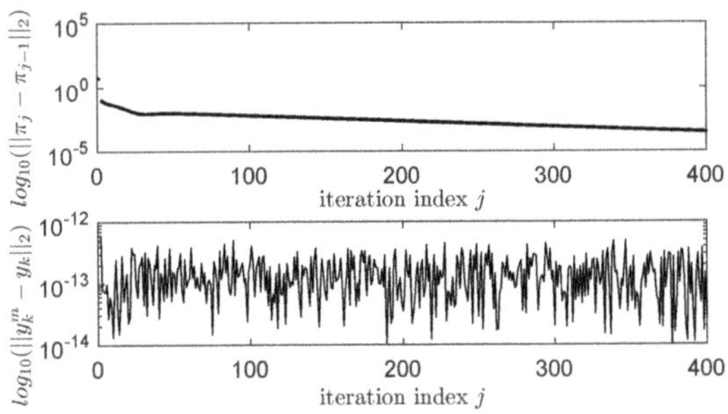

Figure 2. Convergence results of the linearly paramaterized iterative model-free approximate Value Iteration (LP-IMF-AVI) for the linear process example.

4.2. IMF-AVI on the Nonlinear TITOAS Aerodynamic System

The ORM tracking problem on the more challenging TITOAS angular position control [50] (Figure 3) is aimed next. The azimuth (horizontal) motion behaves as an integrator while the pitch (vertical) positioning is affected differently by the gravity for the up and down motions. Coupling between the two channels is present. A simplified deterministic continuous-time state-space model of this process is given as two coupled state-space sub-systems [12,46]:

$$\begin{cases} \dot{\omega}_h = (sat(U_h) - M_h(\omega_h))/2.5 \cdot 10^{-5}, \\ \dot{K}_h = 0.216F_h(\omega_h)cos(\alpha_v) - 0.058\Omega_h + 0.0178sat(U_v)cos(\alpha_v), \\ \Omega_h = K_k/(0.0238cos^2(\alpha_v) + 3 \cdot 10^{-3}), \\ \dot{\alpha}_h = \Omega_h, \\ \dot{\omega}_v = (sat(U_v) - M_v(\omega_v))/1.63 \cdot 10^{-4}, \\ \dot{\Omega}_v = \frac{1}{0.03}\begin{pmatrix} 0.2F_v(\omega_v) - 0.0127\Omega_v - 0.0935sin\alpha_v + \\ -9.28 \cdot 10^{-6}\Omega_v|\omega_v| + 4.17 \cdot 10^{03}sat(U_h) - 0.05cos\alpha_v + \\ -0.021\Omega_h^2 sin\alpha_v cos\alpha_v - 0.093sin\alpha_v + 0.05 \end{pmatrix} \\ \dot{\alpha}_v = \Omega_v, \end{cases} \quad (33)$$

where $sat()$ is the saturation function on $[-1;1]$, $U_h = u_1$ is the azimuth motion control input, $U_v = u_2$ is the vertical motion control input, $\alpha_h(rad) = y_1 \in [-\pi, \pi]$ is the azimuth angle output, $\alpha_v(rad) = y_2 \in [-\pi/2, \pi/2]$ is the pitch angle output, other states being described in [11,48]. The nonlinear static characteristics obtained by polynomial fitting from experimental data are for $\omega_v, \omega_h \in (-4000; 4000)$ [46]:

$$\begin{aligned} M_v(\omega_v) &= 9.05 \times 10^{-12}\omega_v^3 + 2.76 \times 10^{-10}\omega_v^2 + 1.25 \times 10^{-4}\omega_v + 1.66 \times 10^{-4}, \\ F_v(\omega_v) &= -1.8 \times 10^{-18}\omega_v^5 - 7.8 \times 10^{-16}\omega_v^4 + 4.1 \times 10^{-11}\omega_v^3 + 2.7 \times 10^{-8}\omega_v^2 \\ &\quad +3.5 \times 10^{-5}\omega_v - 0.014, \\ M_h(\omega_h) &= 5.95 \times 10^{-13}\omega_h^3 - 5.05 \times 10^{-10}\omega_h^2 + 1.02 \times 10^{-4}\omega_h + 1.61 \times 10^{-3}, \\ F_h(\omega_h) &= -2.56 \times 0^{-20}\omega_h^5 + 4.09 \times 10^{-17}\omega_h^4 + 3.16 \times 10^{-12}\omega_h^3 - 7.34 \times 10^{-9}\omega_h^2 \\ &\quad +2.12 \times 10^{-5}\omega_h + 9.13 \times 10^{-3}. \end{aligned} \quad (34)$$

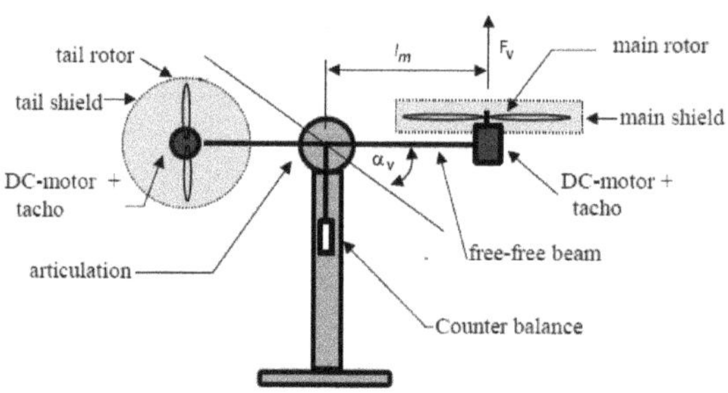

Figure 3. The two-inputs–two-outputs aerodynamic system (TITOAS) experimental setup [50].

A zero-order hold on the inputs and a sampler on the outputs of (33) lead to an equivalent MP discrete-time model of sampling time $T_s = 0.1s$ and of relative degree 1 (one), suitable for input-state data collection

$$P: \begin{cases} \mathbf{x}_{k+1} = \mathbf{f}(\mathbf{x}_k, \mathbf{u}_k), \\ \mathbf{y}_k = \mathbf{g}(\mathbf{x}_k) = [\alpha_{h,k}, \alpha_{v,k}]^\top, \end{cases} \quad (35)$$

where $\mathbf{x}_k = [\omega_{h,k}, \Omega_{h,k}, \alpha_{h,k}, \omega_{v,k}, \Omega_{v,k}, \alpha_{v,k}]^\top \in \mathbb{R}^6$ and $\mathbf{u}_k = [u_{k,1}, u_{k,2}]^\top \in \mathbb{R}^2$. The process' dynamics will not be used for learning the control in the following.

4.3. Initial Controller with Model-Free VRFT

An initial model-free multivariable IO controller is first found using model-free VRFT, as described in [11,24,32]. The ORM is $\mathbf{M}(z) = diag(M_1(z), M_2(z))$ where $M_1(z), M_2(z)$ are the discrete-time counterparts of $M_1(s) = M_2(s) = 1/(3s+1)$ obtained for a sampling period of $T_s = 0.1$ s. The VRFT prefilter is chosen as $\mathbf{L}(z) = \mathbf{M}(z)$. A pseudo-random binary signal (PRBS) of amplitude $[-0.1; 0.1]$ is used on both inputs $u_{k,1}, u_{k,2}$ to open-loop excite the pitch and azimuth dynamics simultaneously, as shown in Figure 4. The IO data $\{\tilde{\mathbf{u}}_k, \tilde{\mathbf{y}}_k\}$ is collected with low-amplitude zero-mean inputs $u_{k,1}, u_{k,2}$, in order to maintain the process linearity around the mechanical equilibrium, such that to fit the linear VRFT design framework (© 2019 IEEE [12]).

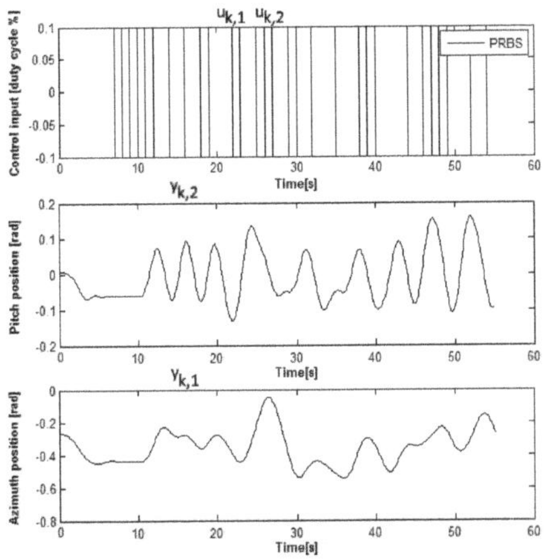

Figure 4. Open-loop input–output (IO) data from the two-inputs–two-outputs aerodynamic system (TITOAS) for Virtual Reference Feedback Tuning (VRFT) controller tuning [46].

An un-decoupling linear output feedback error diagonal controller with the parameters computed by the VRFT approach is [12,46]

$$\begin{aligned} &\mathbf{C}(z,\boldsymbol{\theta}) = \begin{bmatrix} P_{11}(z)/(1-z^{-1}) & 0 \\ 0 & P_{22}(z)/(1-z^{-1}) \end{bmatrix}, \\ &P_{11}(z) = 2.9341 - 5.8689z^{-1} + 3.9303z^{-2} - 0.9173z^{-3} - 0.0777z^{-4}, \\ &P_{22}(z) = 0.6228 - 1.1540z^{-1} + 0.5467z^{-2}, \end{aligned} \tag{36}$$

where the parameter vector $\boldsymbol{\theta}$ groups all the coefficients of $P_{11}(z), P_{22}(z)$. Controller (36) is obtained for $\boldsymbol{\theta}$ as the least squares minimizer of $J_{VR}(\boldsymbol{\theta}) = \sum_{k=1}^{N} \|\tilde{\mathbf{u}}_k^L - \mathbf{C}(z,\boldsymbol{\theta})\tilde{\mathbf{e}}_k^L\|_2^2$ where $\tilde{\mathbf{u}}_k^L = \mathbf{L}(z)\tilde{\mathbf{u}}_k = \mathbf{L}(z)[\tilde{u}_{k,1}, \tilde{u}_{k,2}]^\top, \tilde{\mathbf{e}}_k^L = \mathbf{L}(z)\tilde{\mathbf{e}}_k = \mathbf{L}(z)[\tilde{e}_{k,1}, \tilde{e}_{k,2}]^\top$, $[\tilde{e}_{k,1}, \tilde{e}_{k,2}]^\top = (\mathbf{M}^{-1}(z) - \mathbf{I}_2)[\tilde{y}_{k,1}, \tilde{y}_{k,2}]^\top$. Here, $J_{VR}(\boldsymbol{\theta})$ is an approximation of the c.f. J_{MR}^{∞} from (5) obtained for $\gamma = 1$. The controller (36) will then close the feedback control loop as in $\mathbf{u}_k = \mathbf{C}(z,\boldsymbol{\theta})(\mathbf{r}_k - \mathbf{y}_k)$.

Notice that, by formulation, the VRFT controller tuning aims to minimize the undiscounted ($\gamma = 1$) J_{MR}^{∞} from (5), but via the output feedback controller (36) that processes the feedback control error $\mathbf{e}_k = \mathbf{r}_k - \mathbf{y}_k$. The same goal to minimize (5) is pursued by the subsequent IMF-AVI design of a state-feedback controller tuning for the extended process. Nonlinear (in particular, linear) state-feedback controllers can also be found by VRFT as shown in [24,32], to serve as initializations for the IMF-AVI, or possibly, even for PoIt-like algorithms. However, should this not be necessary, IO feedback controllers are much more data-efficient, requiring significantly less IO data to obtain stabilizing controllers.

4.4. Input–State–Output Data Collection

ORM tracking is intended by making the closed loop CS match the same ORM $\mathbf{M}(z) = diag(M_1(z), M_2(z))$. With the linear controller (36) used in closed-loop to stabilize the process, input–state–output data is collected for 7000 s. The reference inputs with amplitudes $r_{k,1} \in [-2;2], r_{k,2} \in [-1.4;1.1]$ model successive steps that switch their amplitudes uniformly random at 17 s and 25 s, respectively. On the outputs $u_{k,1}, u_{k,2}$ of both controllers $C_{11}(z), C_{22}(z)$, an additive noise is added at every 2nd sample as an uniform random number in $[-1.6;1.6]$ for $C_{11}(z)$ and in $[-1.7;1.7]$ for $C_{22}(z)$. These additive disturbances provide an appropriate exploration, visiting many combinations of input–states–outputs. The computed controller outputs are saturated to $[-1;1]$, then sent to the process. The reference inputs $r_{k,1}, r_{k,2}$ drive the ORM [12,46]:

$$\begin{cases} x^m_{k+1,1} = 0.9672x^m_{k,1} + 0.03278r_{k,1}, \\ x^m_{k+1,2} = 0.9672x^m_{k,2} + 0.03278r_{k,2}, \\ \mathbf{y}^m_k = [y^m_{k,1}, y^m_{k,2}]^\top = [x^m_{k,1}, x^m_{k,2}]^\top. \end{cases} \tag{37}$$

Then the states of the ORM (also outputs of the ORM) are also collected along with the states and control inputs of the process, to build the process extended state (4). Let the extended state be:

$$\mathbf{x}^E_k = [\underbrace{x^m_{k,1}, x^m_{k,2}}_{(\mathbf{x}^m_k)^\top}, \underbrace{r_{k,1}, r_{k,2}}_{\mathbf{r}^\top_k}, (\mathbf{x}_k)^\top]^\top. \tag{38}$$

Essentially, the collected $\mathbf{x}^E_k$ and $\mathbf{u}_k$ builds the transitions dataset $D = \{(\mathbf{x}^E_1, \mathbf{u}_1, \mathbf{x}^E_2), ..., (\mathbf{x}^E_{70000}, \mathbf{u}_{70000}, \mathbf{x}^E_{70001})\}$ for $N = 70{,}000$, used for the IMF-AVI implementation. After collection, an important processing step is the data normalization. Some process states are scaled in order to ensure that all states are inside $[-1;1]$. The scaled process state is $\tilde{\mathbf{x}}_k = [\omega_{h,k}/7200, 25 \cdot \Omega_{h,k}, \alpha_{h,k}, \omega_{v,k}/3500, 40 \cdot \Omega_{v,k}, \alpha_{v,k}]^\top \in \mathbb{R}^6$ and $\mathbf{u}_k = [u_{k,1}, u_{k,2}]^\top \in \mathbb{R}^2$. Other variables such as the reference inputs, the ORM states and the saturated process inputs already have values inside $[-1;1]$. The normalized state is eventually used for state feedback. Collected transition samples are shown in Figure 5 only for the process inputs and outputs, ORM's outputs and reference inputs, for the first 400 s (4000 samples) out of 7000 s [12].

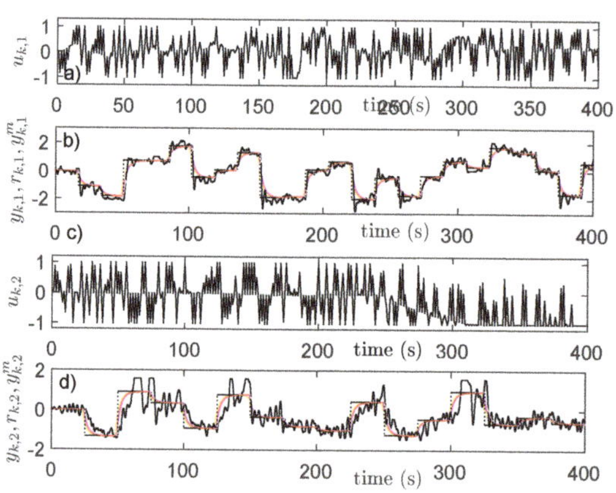

Figure 5. IO data collection with the linear controller: (**a**) $u_{k,1}$; (**b**) $y_{k,1}$ (black), $y^m_{k,1}$ (red), $r_{k,1}$ (black dotted); (**c**) $u_{k,2}$; (**d**) $y_{k,2}$ (black), $y^m_{k,2}$ (red), $r_{k,1}$ (black dotted).

Note that the reference input signals $r_{k,1}, r_{k,2}$ used as sequences of constant amplitude steps for ensuring good exploration, do not have a generative model that obeys the Markov assumption. To avoid this problem, the piece-wise constant reference input generative model $\mathbf{r}_{k+1} = \mathbf{r}_k$ is employed by eliminating from the dataset D all the transition samples that correspond to switching reference inputs instants (i.e., when at least one of $r_{k,1}, r_{k,2}$ switches) (© 2019 IEEE [12]).

4.5. Learning State-Feedback Controllers with Linearly Parameterized IMF-AVI

Details of the LP-IMF-AVI applied to the ORM tracking control problem are next provided. The stage cost is defined $v(\mathbf{x}_k^E) = (y_{k,1} - y_{k,1}^m)^2 + (y_{k,2} - y_{k,2}^m)2$ and the discount factor in J_{MR}^{∞} is $\gamma = 0.95$. The Q-function is linearly parameterized using the basis functions [12]

$$\begin{aligned}\mathbf{\Phi}^\top(\mathbf{x}_k^E, \mathbf{u}_k) =& [(x_{k,1}^m)^2, (x_{k,2}^m)^2, r_{k,1}^2, ..., x_{k,6}^2, u_{k,1}^2, u_{k,2}^2, x_{k,1}^m x_{k,2}^m, x_{k,1}^m r_{k,1}, ...,\\ & x_{k,1}^m u_{k,2}, x_{k,2}^m r_{k,1}, ..., u_{k,1}u_{k,2}] \in \mathbb{R}^{78}.\end{aligned} \tag{39}$$

This basis functions selection is inspired by the shape of the quadratic Q-function resulting from LTI processes with LQR-like penalties (see *Comment 7*). It is expected to be a sensible choice since the TITOAS process is a nonlinear one, therefore the quadratic Q-function may under-parameterize the true Q-function. Nevertheless, its computational advantage incentives the testing of such a solution. Notice that the controller improvement step at each iteration of the LP-IMF-AVI is based on explicit minimization of the Q-function. Solving the linear system of equations resulting after setting the derivative of $Q(\mathbf{x}_k^E, \mathbf{u}_k)$ w.r.t. $\mathbf{u}_k$ equal to zero, it is obtained that (© 2019 IEEE [12])

$$\begin{aligned}&\tilde{\mathbf{u}}_k^* = \begin{bmatrix} u_{k,1}^* \\ u_{k,2}^* \end{bmatrix} = \tilde{\mathbf{C}}^*(\mathbf{x}_k^E, \boldsymbol{\pi}_j) = \begin{bmatrix} 2\pi_{j,11} & \pi_{j,78} \\ \pi_{j,78} & 2\pi_{j,12} \end{bmatrix}^{-1} \begin{bmatrix} F_1(\mathbf{x}_k^E) \\ F_2(\mathbf{x}_k^E) \end{bmatrix},\\ &F_1(\mathbf{x}_k^E) = \pi_{j,22}x_{k,1}^m + \pi_{j,32}x_{k,2}^m + \pi_{j,41}r_{k,1} + \pi_{j,49}r_{k,2} + \pi_{j,56}x_{k,1} +\\ &\quad \pi_{j,62}x_{k,2} + \pi_{j,67}x_{k,3} + \pi_{j,71}x_{k,4} + \pi_{j,74}x_{k,5} + \pi_{j,76}x_{k,6} = \boldsymbol{\pi}_{j,1}^\top \mathbf{x}_k^E,\\ &F_2(\mathbf{x}_k^E) = \pi_{j,23}x_{k,1}^m + \pi_{j,33}x_{k,2}^m + \pi_{j,42}r_{k,1} + \pi_{j,50}r_{k,2} + \pi_{j,57}x_{k,1} +\\ &\quad \pi_{j,63}x_{k,2} + \pi_{j,68}x_{k,3} + \pi_{j,72}x_{k,4} + \pi_{j,75}x_{k,5} + \pi_{j,77}x_{k,6} = \boldsymbol{\pi}_{j,2}^\top \mathbf{x}_k^E.\end{aligned} \tag{40}$$

The improved controller is embedded in the system (12) of 70,000 linear equations with 78 unknowns corresponding to the parameters of $\boldsymbol{\pi}_{j+1} \in \mathbb{R}^{78}$. This linear system (12) is solved in least squares sense, with each of the 50 iterations of the LP-IMF-AVI. The practical convergence results are shown in Figure 6 for $\|\boldsymbol{\pi}_{j+1} - \boldsymbol{\pi}_j\|_2$ and for the ORM tracking performance in terms of a normalized c.f. $J_{test} = 1/N(\|y_{k,1} - y_{k,1}^m\|_2 + \|y_{k,2} - y_{k,2}^m\|_2)$ measured for samples over 200 s in the test scenario displayed in Figure 7. The test scenario consists of a sequence of piece-wise constant reference inputs that switch at different moments of time for the azimuth and pitch ($y_{k,1}$ and $y_{k,2}$, respectively), to illustrate the existing coupling behavior between the two control channels and the extent to which the learned controller manages to achieve the decoupled behavior requested but the diagonal ORM (© 2019 IEEE [12]).

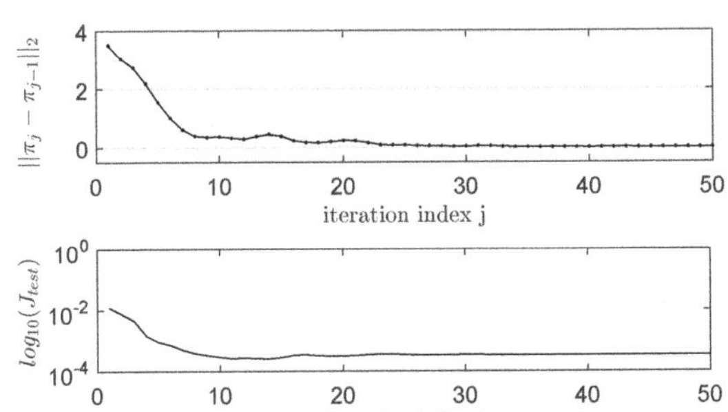

Figure 6. The LP-IMF-AVI convergence on TITOAS (© 2019 IEEE [12]).

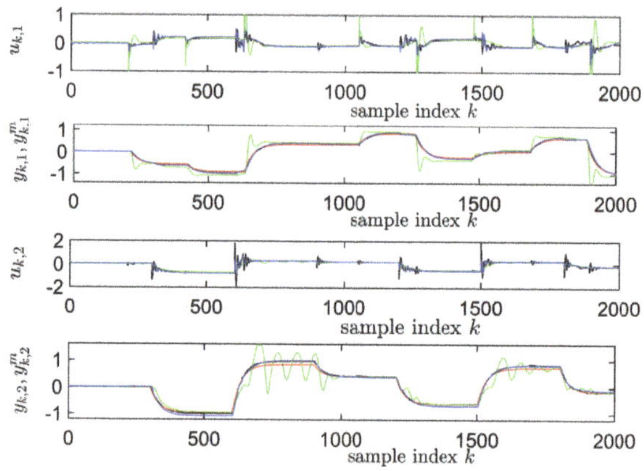

Figure 7. The IMF-AVI convergence on TITOAS: $y^m_{k,1}, y^m_{k,2}$, (red); $u_{k,1}, u_{k,2}, y_{k,1}, y_{k,2}$ for LP-IMF-AVI (black), for NP-IMF-AVI with NNs (blue), for the initial VRFT controller used for transitions collection (green) (© 2019 IEEE [12]).

The best LP-IMF-AVI controller found over the 50 iterations results in $J_{test} = 0.0017$ (tracking results in black lines in Figure 7), which is more than 6 times lower than the tracking performance of the VRFT controller used for transition samples collection, for which $J_{test} = 0.0103$ (tracking results in green lines in Figure 7) (© 2019 IEEE [12]). The convergence of the LP-IMF-AVI parameters is depicted in Figure 8.

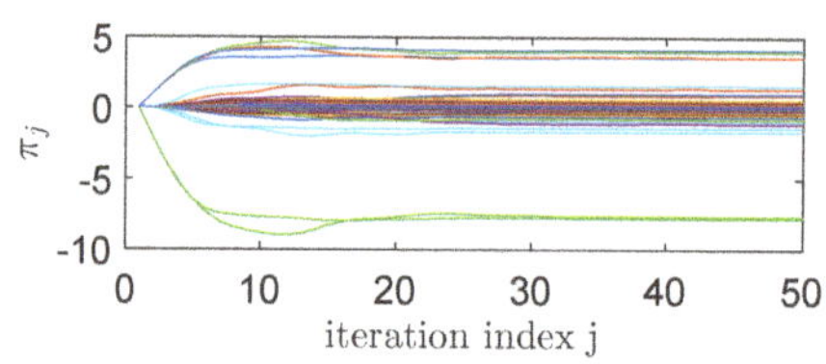

Figure 8. The LP-IMF-AVI parameters convergence.

4.6. Learning State-Feedback Controllers with Nonlinearly Parameterized IMF-AVI Using NNs

The previous LP-IMF-AVI for ORM tracking control learning scheme is next challenged by a NP-IMF-AVI implemented with NNs. In this case, two NNs are used to approximate the Q-function and the controller (the latter is sometimes avoidable, see the comments later on in this sub-section). The procedure follows the NP-IMF-AVI implementation described in [24,51]. The same dataset of transition samples is used as was previously used for the LP-IMF-AVI. Notice that the NN-based implementation is widely used in the reinforcement learning-based approach of ADP and is generally more scalable to problems of high dimension.

The controller NN (C-NN) estimate is a 10–3–2 (10 inputs because $\mathbf{x}^E_k \in \mathbb{R}^{10}$, 3 neurons in the hidden layer, and 2 outputs corresponding to $u_{k,1}, u_{k,2}$) with *tanh* activation function in the hidden layer and linear output activation. The Q-function NN (Q-NN) estimate is 12–25–1 with the same parameters as C-NN. Initial weights of both NNs are uniform random numbers with zero-mean and variance 0.3. Both NNs are to be trained using scaled conjugate gradient for a maximum of 500 epochs. The available dataset is randomly divided into training (80%) and validation data (20%). Early stopping during training is enforced after 10 increases of the training c.f. mean sum of squared errors (MSE) evaluated on the validation data. MSE is herein, for all networks, the default performance function used in training (© 2019 IEEE [12]).

The NP-IMF-AVI proposed herein consists of two steps for each iteration j. The first one calculates the targets for the NN $Q(\mathbf{x}_k^E, \mathbf{u}_k, \boldsymbol{\pi}_j)$ (having inputs $[(\mathbf{x}_k^E)^\top, (\mathbf{u}_k)^\top]^\top$ and current iteration weights $\boldsymbol{\pi}_j$) as $\{v(\mathbf{x}_k^E, \mathbf{u}_k) + \gamma Q(\mathbf{x}_{k+1}^E, \mathbf{C}(\mathbf{x}_{k+1}^E, \boldsymbol{\theta}_{j-1}), \boldsymbol{\pi}_{j-1})\}$, for all transitions in the dataset. Resulting in the trained Q-function estimator NN $Q(\mathbf{x}_k^E, \mathbf{u}_k, \boldsymbol{\pi}_j)$ with parameter weights $\boldsymbol{\pi}_j$. The second step (the controller improvement) first calculates the targets for the controller $\mathbf{C}(\mathbf{x}_k^E, \boldsymbol{\theta}_j)$ (with inputs $(\mathbf{x}_k^E)^\top$) as $\{\mathbf{u}_k = argmin_{\mathbf{u}\in\Lambda} Q(\mathbf{x}_k^E, \mathbf{u}, \boldsymbol{\pi}_j)\}$. Note that additional parameterization for the controller NN weights $\boldsymbol{\theta}_j$ is needed. Training produces the improved controller characterized by the new weights $\boldsymbol{\theta}_j$. Here, the discrete set of control actions $\Lambda \subset \Omega_U$ used to minimize the Q-NN estimate for computing the controller targets is the Cartesian product of two identical sets of control actions, each containing 21 equally spaced values in [–1;1], i.e., $\{-1, -0.9, ..., 0.9, 1\}$.

A discount $\gamma = 0.95$ will be used and each iteration of the NP-IMF-AVI produces a C-NN that is tested on the standard test scenario shown in Figure 6 by measuring the same normalized c.f. J_{test} for $N = 2000$, on the same test scenario that was used in the case of the LP-IMF-AVI. The NP-IMF-AVI is iterated 50 times and all the stabilizing controllers that are better than the VRFT multivariable controller running on the standard test scenario described in Figure 7 (in terms of smaller J_{test}) are stored. The best C-NN across 50 iterations renders $J_{test} = 0.0025$. The tracking performance for the best NN controller found with the NP-IMF-AVI is shown in blue lines in Figure 7. The convergence process is depicted in Figure 9.

A gridsearch is next performed for the NP-IMF-AVI training process, by changing the dataset size from 30,000 to 50,000 to 70,000, combined with 17, 19, and 21 discrete values used for minimizing the Q-function over the two control inputs. For the case of 50,000 data with 17 uniform discrete possible values for each control input, $J_{test} = 0.0017$ which is the same with the best performance of the LP-IMF-AVI. Notice that neither the nonlinear state-feedback controller of the NP-IMF-AVI nor the linear state-feedback controller of the LP-IMF-AVI have integral component, while the linear output feedback controller tuned by VRFT and used for exploration has integrators.

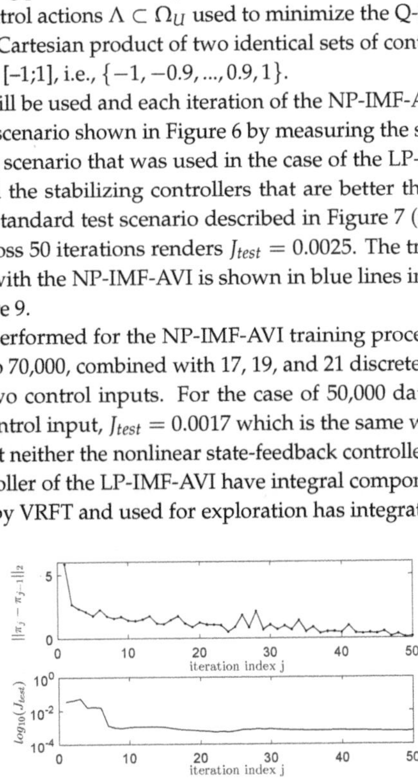

Figure 9. The nonlinearly parameterized IMF-AVI (NP-IMF-AVI) convergence.

Two additional approaches exist for dealing with a NP-IMF-AVI using two NNs, for each of the Q-NN approximator and for the C-NN. For example, [32] used to cascade the C-NN and the Q-NN. After training the Q-NN and producing the new weights $\boldsymbol{\pi}_j$, the weights of the Q-NN are fixed and only the weights $\boldsymbol{\theta}_j$ of the C-NN are trained, with all the targets equal to $\{0\}$ for all the inputs $\mathbf{x}_k^E$ of the cascaded NN $Q(\mathbf{x}_k^E, \mathbf{C}(\mathbf{x}_k^E, \boldsymbol{\theta}_j), \boldsymbol{\pi}_j)$. In this way, the C-NN is forced to minimize the Q-NN. The disadvantage is the vanishing gradient problem of the resulted cascaded network that deepens through more hidden layers, therefore only small corrections are brought to the C-NN part that is further away from the Q-NN's output. Yet another solution [14] uses, for the controller improvement step, a single/several gradient descent step/steps $\boldsymbol{\theta}_j = \boldsymbol{\theta}_{j-1} - \alpha \frac{1}{N}\sum_{k=1}^{N} \frac{dQ}{d\mathbf{u}}\frac{d\mathbf{u}}{d\boldsymbol{\theta}}|_{(\boldsymbol{\theta}_{j-1},\mathbf{x}_k^E)}$ with each iteration of NP-IMF-AVI, with step size $\alpha > 0$ and with gradient $\frac{1}{N}\sum_{k=1}^{N} \frac{dQ}{d\mathbf{u}}\frac{d\mathbf{u}}{d\boldsymbol{\theta}}|_{(\boldsymbol{\theta}_{j-1},\mathbf{x}_k^E)}$ accumulated over all inputs $\mathbf{x}_k^E$ of the cascaded NN $Q(\mathbf{x}_k^E, \mathbf{C}(\mathbf{x}_k^E, \boldsymbol{\theta}_{j-1}), \boldsymbol{\pi}_j)$, over fixed Q-NN weights. Essentially, the two approaches described above are equivalent and the number of gradient descent steps at each iteration is user-selectable. Also, no minimization by enumerating a finite set of control actions needs to be performed in either of the two above approaches. The above two equivalent approaches are effectively a particular case of the Neural-Fitted Q-iteration with Continuous Actions (NFQCA) approach [14], more recently to be updated with some changes to Deep Deterministic Policy Gradient (DDPG) [52]. DDPG uses two NNs as well, for the Q-NN and for the C-NN. It was originally developed to work in online off-policy mode, hence the need to update the Q-NN and the C-NN in a faster way on a relatively small number of transition samples (called minibatch) randomly extracted from a replay buffer equivalent to the dataset D, in order to break the time correlation of consecutive samples. The effectiveness of DDPG in real-time online control has yet to be proven.

Two variants of *offline* off-policy DDPG called DDPG1 and DDPG2 are run for comparisons purposes. Both use minibatches of 128 transitions from the dataset D at each training iteration. While both use soft target updates of the Q-NN weights $\pi_j' = \tau\pi_j + (1-\tau)\pi_{j-1}'$ and of the C-NN weights $\theta_j' = \tau\theta_j + (1-\tau)\theta_{j-1}'$, with $\tau = 0.005$. π_j' and θ_j' are used to calculate the targets for the Q-NN training. At each iteration, DDPG1 makes one update step of the Q-NN weights in the negative direction of the gradient of the MSE w.r.t. π_j with step size $\alpha = 0.001$ and one update step of the C-NN weights in the negative direction of the gradient of the Q-NN's output w.r.t. θ_j with step size $\alpha = 0.001$. While DDPG2 differs in that the Q-NN training on each minibatch of each iteration is left to the same settings used for NP-IMF-AVI training (scaled conjugate gradient for maximum 500 epochs), only one gradient descent step is used to update the C-NN weights with the same $\alpha = 0.001$. The step-sizes were selected to ensure learning convergence. It was observed that DDPG1 has the slowest convergence (convergence appears after more than 20,000 iterations) since it performs only one gradient update step per iteration, DDPG2 has faster convergence speed (convergence appears after 5000 iterations) since it allows more gradient steps for Q-NN training, while NP-IMF-AVI has the highest convergence speed (convergence appears after 10 iterations), allowing more training in terms of gradient descent steps (with scaled conjugate gradient direction) for both Q-NN and for C-NN, at each iteration. This proves that, given the high-dimensional process, it is better to use the entire dataset D for offline training, as it was done with NP-IMF-AVI. On the other hand, the best performance with DDPG1 and DDPG2 is 0.003, not as good as the best one with the more computationally demanding NP-IF-AVI (0.0017), suggesting that minimizing the Q-NN by enumerating discrete actions to calculate the C-NN targets may actually escape local minima. The total learning time to convergence with DDPG1 and DDPG2 is about the same as with NP-IMF-AVI, which is to be expected since less calculations for DDPG1 takes more iterations until convergence appears. Notice that NP-IMF-AVI does not use soft target updates for its two NNs.

The additional NN controller is not mandatory and the NP-IMF-AVI can be made similar to the LP-IMF-AVI case. In this case, the minimization of the Q-function NN estimate is to be performed by enumerating the discrete set of control actions $\Lambda \subset \Omega_U$ and the targets calculation for the Q-function NN will use $\{v(\mathbf{x}_k^E, \mathbf{u}_k) + \gamma Q(x_{k+1}^E, argmin_{\mathbf{u}\in\Lambda} Q(\mathbf{x}_{k+1}^E, \mathbf{u}, \pi_j), \pi_{j-1})\}$. This approach merges the controller improvement step and the Q-function improvement step. However, for real-time control implementation after NP-IMF-AVI convergence, it is more expensive to find $\mathbf{u}_k^* = argmin_{\mathbf{u}\in\Lambda} Q^*(\mathbf{x}_k^E, \mathbf{u}, \pi_j)$, since it requires evaluating the Q-function NN for a number of times proportional to the number of combinations of discrete control actions. Then only slower processes can be accommodated with this implementation. Whereas in the case when a dedicated controller NN is used, after NP-IMF-AVI convergence, the optimal control $\mathbf{u}_k^* = \mathbf{C}^*(\mathbf{x}_k^E, \theta_j)$ is calculated at once, through a single NN evaluation. This dedicated NN controller can also be obtained (trained) as a final step after the NP-IMF-AVI has converged to the optimal Q-function $Q^*(\mathbf{x}_k^E, \mathbf{u}, \pi_j)$ and the targets for the controller output are calculated as $\{\mathbf{u}_k^* = argmin_{\mathbf{u}\in\Lambda} Q^*(\mathbf{x}_k^E, \mathbf{u}, \pi_j)\}$. Another original solution that uses a single NN Q-function approximator was proposed in [53], such that a quadratic approximation of the NN-fitted Q-function is used to directly derive a linear state-feedback controller with each iteration.

4.7. Comments on the Obtained Results

Some comments follow the validation of the LP-IMF-AVI and NP-IMF-AVI. The results of Figure 6 indicate that convergence of the LP-IMF-AVI is attained in terms of $\|\pi_j - \pi_{j-1}\|_2 \to 0$, however perfect ORM tracking is not possible, as shown by the nonzero constant value of J_{test}. On one hand, this is to be expected since the resulting linear state-feedback controller coupled with the process' nonlinear dynamics is not capable of ensuring a closed-loop linear behavior as requested by the ORM. On the other hand, the NN controller resulting from the NP-IMF-AVI implementation is a nonlinear state feedback controller, however the best obtained results are not better than (but on the same level with) those obtained with the linear state-feedback controller of the NP-IMF-AVI, although the nonlinear controller is expected to perform better in terms of lower J_{test}, due to its flexibility being able to compensate for the process nonlinearity. If this flexibility does not turn into an advantage, the reason lies with the additional NN controller parameterization (that introduces additional approximation errors) and with the training process that relies on approximate minimizations in the calculation step of the controller's targets.

The iterative evolution of J_{test} in case of both LP-IMF-AVI and NP-IMF-AVI show stabilization to constant nonzero values, suggesting that neither approach can provide perfectly ORM tracking controllers. For the LP-IMF-AVI, the responsibility lies with the under-representation error introduced by quadratic Q-function (and with the subsequent resulting linear state-feedback controller), while for NP-IMF-AVI, responsibility lies with the errors introduced by the additional controller approximator NN and the targets calculation in the controller improvement step.

Computational resources analysis indicate that the LP-IMF-AVI has learned only 78 parameters for to the Q-function parameterization, and no intermediate controller approximator is used. The run time for 50 iterations is about 345 seconds (including evaluation steps on the test scenario after each iteration). The NN-based NP-IMF-AVI needs to learn two NNs having 351 parameters (weights) for the Q-function NN and 41 parameters (weights) for the controller NN, respectively. Contrastingly, the runtime for the NP-IMF-AVI is about 3300 seconds, almost ten times more than in the case of the LP-IMF-AVI. Despite the larger parameter learning space, the converged behavior of the NN-based NP-IMF-AVI is very similar to that of the LP-IMF-AVI (see tracking results in Figure 7).

LP-IMF-AVI has shown an increased sensitivity to the transition samples dataset size: for fewer or more transition samples in the dataset, the LP-IMF-AVI diverges, under exactly the same exploration settings. But this divergence appears only after an initial convergence phase similar to that of Figure 6, and not from the very beginning. Whereas having fewer transition samples is intuitively disadvantageous for learning the true Q-function approximation, having a larger number of transition samples leading to divergence is unexpected. The reason is that non-uniform state-action space exploration affects the linear regression. Then, given a fixed dataset size, an increased amplitude of the additive disturbance used to stimulate exploration combined with a more often application of this disturbance (such as every 2nd sample) increases the convergence probability. These observations indicate again that the proposed linear parameterization using quadratic basis functions is insufficient for a correct representation of the true Q-function, thus failing the small approximation errors assumptions of Theorem 2. The connection between the convergence guarantees and the approximation errors have been analyzed in the literature [54–57].

In the light of the previous paragraph's observations, the NN-based NP-IMF-AVI proves to be significantly more robust throughout the convergence process, both to various transition samples dataset sizes and to different exploration settings (disturbance amplitude and frequency of its application, how often the reference inputs switch during the transition samples collection phase, etc.). This may well pay off for the additional controller approximator NN and for the extra computation time since the chances of learning high performance controllers will depend less on the selection of the many parameters involved. Moreover, manual selection of the basis functions is unnecessary with the NN-based NP-IMF-AVI, while the over-parameterization is automatically managed by the NN training mechanism.

Data normalization is a frequently overlooked issue in ADP control but it is critical to successful design since it numerically affects both the regression solution in LP-IMF-AVI and the NN training in NP-IMF-AVI. A diagonal scaling matrix $\mathbf{S} = diag(s_1, ..., s_{n+n_m+p})$ leads to the scaled extended state $\bar{\mathbf{x}}_k^E = \mathbf{S}\mathbf{x}_k^E$ resulting in the extended state-space model $\bar{\mathbf{x}}_{k+1}^E = \mathbf{S} \cdot \mathbf{E}(\mathbf{S}^{-1}\bar{\mathbf{x}}_k^E, \mathbf{u}_k)$ that still preserves the MDP property.

5. Conclusions

This paper proves a functional design for an IMF-AVI ADP learning scheme dedicated to the challenging problem of ORM tracking control for a high-order real-world complex nonlinear process with unknown dynamics. The investigation revolves around a comparative analysis of a linear vs. a nonlinear parameterization of the IMF-AVI approach. Learning high performance state-feedback control under the model-free mechanism offered by IMF-AVI builds upon the input–states–outputs transition samples collection step that uses an initial exploratory linear output feedback controller that is also designed in a model-free setup using VRFT. From the practitioners' viewpoint, the NN-based implementation of IMF-AVI is more appealing since it easily scales up with problem dimension and automatically manages the basis functions selection for the function approximators.

Future work attempts to validate the proposed design approach to more complex high-order nonlinear processes of practical importance.

Author Contributions: conceptualization, M.-B.R.; methodology, M.-B.R.; software, T.L; validation, T.L.; formal analysis, M.-B.R.; investigation, T.L.; data curation, M.-B.R. and T.L.; writing–original draft preparation, M.-B.R. and T.L.; writing–review and editing, M.-B.R. and T.L.; supervision, M.-B.R.;

Funding: This research received no external funding

Conflicts of Interest: The authors declare no conflict of interest.

References

1. Radac, M.B.; Precup, R.E.; Petriu, E.M. Model-free primitive-based iterative learning control approach to trajectory tracking of MIMO systems with experimental validation. *IEEE Trans. Neural Netw. Learn. Syst.* **2015**, *26*, 2925–2938. [CrossRef]
2. Sutton, R.S.; Barto, A.G. *In Reinforcement Learning: An Introduction*; MIT Press: Cambridge, MA, USA, 1998.
3. Bertsekas, D.P.; Tsitsiklis, J.N. In *Neuro-Dynamic Programming*; Athena Scientific: Belmont, MA, USA, 1996.
4. Wang, F.Y.; Zhang, H.; Liu, D. Adaptive dynamic programming: an introduction. *IEEE Comput. Intell. Mag.* **2009**, *4*, 39–47. [CrossRef]
5. Lewis, F.; Vrabie, D.; Vamvoudakis, K.G. Reinforcement learning and feedback control: Using natural decision methods to design optimal adaptive controllers. *IEEE Control Syst. Mag.* **2012**, *32*, 76–105.
6. Lewis, F.; Vrabie, D.; Vamvoudakis, K.G. Reinforcement learning and adaptive dynamic programming for feedback control. *IEEE Circ. Syst. Mag.* **2009**, *9*, 76–105. [CrossRef]
7. Murray, J.; Cox, C.J.; Lendaris, G.G.; Saeks, R. Adaptive dynamic programming. *IEEE Trans. Syst. Man Cybern.* **2002**, *32*, 140–153. [CrossRef]
8. Mnih, V.; Kavukcuoglu, K.; Silver, D.; Rusu, A.A.; Veness, J.; Bellemare, M.G.; Graves, A.; Riedmiller, M.; Fidjeland, A.K.; Ostrovski, G.; et al. Human-level control through deep reinforcement learning. *Nature* **2015**, *518*, 529–533. [CrossRef]
9. Kiumarsi, B.; Lewis, F.L.; Naghibi-Sistani, M.B.; Karimpour, A. Optimal tracking control of unknown discrete-time linear systems using input–output measured data. *IEEE Trans. Cybern.* **2015**, *45*, 2270–2779. [CrossRef]
10. Kiumarsi, B.; Lewis, F.L.; Modares, H.; Karimpour, A.; Naghibi-Sistani, M.B. Reinforcement Q-learning for optimal tracking control of linear discrete-time systems with unknown dynamics. *Automatica* **2014**, *50*, 1167–1175. [CrossRef]
11. Radac, M.B.; Precup, R.E.; Roman, R.C. Model-free control performance improvement using virtual reference feedback tuning and reinforcement Q-learning. *Int. J. Syst. Sci.* **2017**, *48*, 1071–1083. [CrossRef]

12. Lala, T.; Radac, M.B. Parameterized value iteration for output reference model tracking of a high order nonlinear aerodynamic system. In Proceedings of the 2019 27th Mediterranean Conference on Control and Automation (MED), Akko, Israel, 1–4 July 2019; pp. 43–49.
13. Ernst, D.; Geurts, P.; Wehenkel, L. Tree-based batch mode reinforcement learning. *J. Mach. Learn. Res.* **2005**, *6*, 2089–2099.
14. Hafner, R.; Riedmiller, M. Reinforcement learning in feedback control. Challenges and benchmarks from technical process control. *Mach. Learn.* **2011**, *84*, 137–169. [CrossRef]
15. Zhao, D.; Wang, B.; Liu, D. A supervised actor critic approach for adaptive cruise control. *Soft Comput.* **2013**, *17*, 2089–2099. [CrossRef]
16. Cui, R.; Yang, R.; Li, Y.; Sharma, S. Adaptive neural network control of AUVs with Control input nonlinearities using reinforcement learning. *IEEE Trans. Syst. Man Cybern.* **2017**, *47*, 1019–1029. [CrossRef]
17. Xu, X.; Hou, Z.; Lian, C.; He, H. Online learning control using adaptive critic designs with sparse kernel machines. *IEEE Trans. Neural Netw. Learn. Syst.* **2013**, *24*, 762–775.
18. He, H.; Ni, Z.; Fu, J. A three-network architecture for on-line learning and optimization based on adaptive dynamic programming. *Neurocomputing* **2012**, *78*, 3–13 [CrossRef]
19. Modares, H.; Lewis, F.L.; Jiang, Z.P. H_∞ Tracking control of completely unknown continuous-time systems via off-policy reinforcement learning. *IEEE Trans. Neural Netw. Learn. Syst.* **2015**, *26*, 2550–2562. [CrossRef]
20. Li, J.; Modares, H.; Chai, T.; Lewis, F.L.; Xie, L. Off-policy reinforcement learning for synchronization in multiagent graphical games. *IEEE Trans. Neural Netw. Learn. Syst.* **2017**, *28*, 2434–2445. [CrossRef]
21. Bertsekas, D. Value and policy iterations in optimal control and adaptive dynamic programming. *IEEE Trans. Neural Netw. Learn. Syst.* **2017**, *28*, 500–509. [CrossRef]
22. Yang, Y.; Wunsch, D.; Yin, Y. Hamiltonian-driven adaptive dynamic programming for continuous nonlinear dynamical systems. *IEEE Trans. Neural Netw. Learn. Syst.* **2017**, *28*, 1929–1940. [CrossRef]
23. Kamalapurkar, R.; Andrews, L.; Walters, P.; Dixon, W.E. Model-based reinforcement learning for infinite horizon approximate optimal tracking. *IEEE Trans. Neural Netw. Learn. Syst.* **2017**, *28*, 753–758. [CrossRef]
24. Radac, M.B.; Precup, R.E. Data-driven MIMO Model-free reference tracking control with nonlinear state-feedback and fractional order controllers. *Appl. Soft Comput.* **2018**, *73*, 992–1003. [CrossRef]
25. Campi, M.C.; Lecchini, A.; Savaresi, S.M. Virtual Reference Feedback Tuning: A direct method for the design of feedback controllers. *Automatica* **2002**, *38*, 1337–1346. [CrossRef]
26. Hjalmarsson, H. Iterative Feedback Tuning—An overview. *Int. J. Adapt. Control Signal Process.* **2002**, *16*, 373–395. [CrossRef]
27. Janssens, P.; Pipeleers, G.; Swevers, J.L. Model-free iterative learning control for LTI systems and experimental validation on a linear motor test setup. In Proceedings of the 2011 American Control Conference (ACC), San Francisco, CA, USA, 29 June–1 July 2011; pp. 4287–4292.
28. Radac, M.B.; Precup, R.E. Optimal behavior prediction using a primitive-based data-driven model-free iterative learning control approach. *Comput. Ind.* **2015**, *74*, 95–109. [CrossRef]
29. Chi, R.; Hou, Z.S.; Jin, S.; Huang, B. An improved data-driven point-to-point ILC using additional on-line control inputs with experimental verification. *IEEE Trans. Syst. Man Cybern.* **2017**, *49*, 687–696. [CrossRef]
30. Abouaissa, H.; Fliess, M.; Join, C. On ramp metering: towards a better understanding of ALINEA via model-free control. *Int. J. Control* **2017**, *90*, 1018–1026. [CrossRef]
31. Hou, Z.S.; Liu, S.; Tian, T. Lazy-learning-based data-driven model-free adaptive predictive control for a class of discrete-time nonlinear systems. *IEEE Trans. Neural Netw. Learn. Syst.* **2017**, *28*, 1914–1928. [CrossRef]
32. Radac, M.B.; Precup, R.E.; Roman, R.C. Data-driven model reference control of MIMO vertical tank systems with model-free VRFT and Q-learning. *ISA Trans.* **2018**, *73*, 227–238. [CrossRef]
33. Bolder, J.; Kleinendorst, S.; Oomen, T. Data-driven multivariable ILC: enhanced performance by eliminating L and Q filters. *Int. J. Robot. Nonlinear Control* **2018**, *28*, 3728–3751. [CrossRef]
34. Wang, Z.; Lu, R.; Gao, F.; Liu, D. An indirect data-driven method for trajectory tracking control of a class of nonlinear discrete-time systems. *IEEE Trans. Ind. Electron.* **2017**, *64*, 4121–4129. [CrossRef]
35. Pandian B.J.; Noel, M.M. Control of a bioreactor using a new partially supervised reinforcement learning algorithm. *J. Proc. Control* **2018**, *69*, 16–29. [CrossRef]
36. Diaz, H.; Armesto, L.; Sala, A. Fitted q-function control methodology based on takagi-sugeno systems. *IEEE Trans. Control Syst. Tech.* **2018**. [CrossRef]

37. Wang, W.; Chen, X.; Fu, H.; Wu, M. Data-driven adaptive dynamic programming for partially observable nonzero-sum games via Q-learning method. *Int. J. Syst. Sci.* **2019**. [CrossRef]
38. Mu, C.; Zhang, Y. Learning-based robust tracking control of quadrotor with time-varying and coupling uncertainties. *IEEE Trans. Neural Netw. Learn. Syst.* **2019**. [CrossRef] [PubMed]
39. Liu, D.; Yang, G.H. Model-free adaptive control design for nonlinear discrete-time processes with reinforcement learning techniques. *Int. J. Syst. Sci.* **2018**, *49*, 2298–2308. [CrossRef]
40. Song, F.; Liu, Y.; Xu, J.X.; Yang, X., Zhu, Q. Data-driven iterative feedforward tuning for a wafer stage: A high-order approach based on instrumental variables. *IEEE Trans. Ind. Electr.* **2019**, *66*, 3106–3116. [CrossRef]
41. Kofinas, P.; Dounis, A. Fuzzy Q-learning agent for online tuning of PID controller for DC motor speed control. *Algorithms* **2018**, *11*, 148. [CrossRef]
42. Radac, M.-B.; Precup, R.-E. Three-level hierarchical model-free learning approach to trajectory tracking control. *Eng. Appl. Artif. Intell.* **2016**, *55*, 103–118. [CrossRef]
43. Radac, M.-B.; Precup, R.-E. Data-based two-degree-of-freedom iterative control approach to constrained non-linear systems. *IET Control Theory Appl.* **2015**, *9*, 1000–1010. [CrossRef]
44. Salgado, M.; Clempner, J.B. Measuring the emotional state among interacting agents: A game theory approach using reinforcement learning. *Expert Syst. Appl.* **2018**, *97*, 266–275. [CrossRef]
45. Silva, M.A.L.; de Souza, S.R.; Souza, M.J.F.; Bazzan, A.L.C. A reinforcement learning-based multi-agent framework applied for solving routing and scheduling problems. *Expert Syst. Appl.* **2019**, *131*, 148–171. [CrossRef]
46. Radac, M.-B; Precup, R.-E.; Hedrea E.-L.; Mituletu I.-C. Data-driven model-free model-reference nonlinear virtual state feedback control from input-output data. In Proceedings of the 2018 26th Mediterranean Conference on Control and Automation (MED), Zadar, Croatia, 19–22 June 2018; pp. 332–338.
47. Campestrini, L.; Eckhard, D.; Gevers, M.; Bazanella, M. Virtual reference tuning for non-minimum phase plants. *Automatica* **2011**, *47*, 1778–1784. [CrossRef]
48. Al-Tamimi, A.; Lewis, F.L.; Abu-Khalaf, M. Discrete-time nonlinear HJB Solution using approximate dynamic programming: Convergence proof. *IEEE Trans. Syst. Man Cybern. Cybern.* **2008**, *38*, 943–949. [CrossRef] [PubMed]
49. Rantzer, A. Relaxed dynamic programming in switching systems. *IEEE Proc. Control Theory Appl.* **2006**, *153*, 567–574. [CrossRef]
50. Inteco, LTD. *Two Rotor Aerodynamical System*; User's Manual; Inteco, LTD: Krakow, Poland, 2007. Available online: http://ee.sharif.edu/~lcsl/lab/Tras_um_PCI.pdf (accessed on 12 June 2019).
51. Radac, M.-B; Precup, R.-E. Data-driven model-free slip control of anti-lock braking systems using reinforcement Q-learning. *Neurocomputing* **2018**, *275*, 317–329. [CrossRef]
52. Lillicrap, T.P.; Hunt, J.J.; Pritzel A.; Heess N.; Erez, T.; Tassa, Y.; Silver, D.; Wierstra, D. Continuous control with deep reinforcement learning. *Mach. Learn.* **2016**, arXiv:1509.02971.
53. Ten Hagen, S.; Krose, B. Neural Q-learning. *Neural Comput. Appl.* **2003**, *12*, 81–88. [CrossRef]
54. Radac, M.B.; Precup, R.E. Data-Driven model-free tracking reinforcement learning control with VRFT-based adaptive actor-critic. *Appl. Sci.* **2019**, *9*, 1807. [CrossRef]
55. Dierks, T.; Jagannathan, S. Online optimal control of affine nonlinear discrete-time systems with unknown internal dynamics by using time-based policy update. *IEEE Trans. Neural Netw. Learn. Syst.* **2012**, *23*, 1118–1129. [CrossRef]
56. Heydari, A. Theoretical and numerical analysis of approximate dynamic programming with approximation errors. *J. Gui Control Dyn.* **2016**, *39*, 301–311. [CrossRef]
57. Heydari, A. Revisiting approximate dynamic programming and its convergence. *IEEE Trans. Cybern.* **2014**, *44*, 2733–2743. [CrossRef] [PubMed]

algorithms

MDPI

Article

Kalman-Filter-Based Tension Control Design for Industrial Roll-to-Roll System

Hyeongjin Hwang †, Jehwon Lee †, Sangjune Eum and Kanghyun Nam *

Graduate School of Mechanical Engineering, Yeungnam University, 280 Daehak-ro, Gyeongsan 38541, Korea; ppidol3@ynu.ac.kr (H.H.); jwlee@yu.ac.kr (J.L.); power1823@naver.com (S.E.)

* Correspondence: khnam@yu.ac.kr; Tel.: +82-53-810-2455

† These authors contributed equally to this work.

Received: 2 April 2019; Accepted: 22 April 2019; Published: 24 April 2019

Abstract: This paper presents a robust and precise tension control method for a roll-to-roll (R2R) system. In R2R processing, robust and precise tension control is very important because improper web tension control leads to deterioration in the quality of web material. However, tension control is not easy because the R2R system has a model variation in which the inertia of the web in roll form is changed and external disturbances caused by web slip and crumpled web. Therefore, a disturbance observer (DOB) was proposed to achieve robustness against model variations and external disturbances. DOB is a robust control method widely used in various fields because of its simple structure and excellent performance. Moreover, the web passes through various process steps to achieve the finished product in the R2R process. Particularly, it is important to track the tension when magnitude of the tension varies during process. Feedforward (FF) controller was applied to minimize the tracking error in the transient section where tension changes. Moreover, the signal processing of a sensor using the Kalman filter (KF) in the R2R system greatly improved control performance. Finally, the effectiveness of the proposed control scheme is discussed using experimental results.

Keywords: disturbance observer; Kalman filter; feedforward control; tension controller; roll-to-roll system

1. Introduction

Roll-to-roll (R2R) technology has recently attracted a great deal of attention for the mass production of electronic devices in the field of display and battery [1]. The R2R machine is composed of rewinder, unwinder, and feeder motors. The rewinder motor is used to wind the web made in roll form with the desired tension. On the other hand, the unwinder motor is used to unwind the web. The feeder motor is used for web transfer motions at the proper speed.

In the R2R system, tension control is usually achieved by using a load cell or dancer roll. In R2R processing, robust and precise tension control is essential because it affects the quality of the web materials. The web has flexible substrates including metal foils, glass, and ceramics. In particular, these substrates used in electronic devices require precise tension control because coating and printing thickness variation occurs despite minute tension changes, resulting in changes in electrical properties.

Some studies have investigated the tension control of R2R equipment. In 1993, Ebler investigated web tension control with dancer rolls and load cells [2]. The two different systems have been analyzed and experimental results have been provided. In 1998, K. Okada designed an adaptive fuzzy control for a web tension control system [3]. In 2002, Koç analyzed modeling and robust control of a winding system [4]. The effectiveness of robust control strategy in a web system is compared to a proportional-integral-derivative (PID) controller commonly used in the industry. In 2007, Shin presented the effect of tension on the lateral dynamics and control of a moving web [5]. Experimental studies of the lateral motion of a web were carried out and a cross-couple controller was proposed, which automatically tunes the proportional and integral gains of a lateral-position controller according to the web tension.

However, despite these numerous studies, tension control is not easy because it has model variations in which the inertia of the web in roll form is changed and external disturbances caused by web slip and crumpled web. To achieve robustness against model variations and external disturbances, robust control such as a disturbance observer (DOB) is required [6–10]. DOB has been widely used as an effective methodology to overcome model uncertainty and external disturbance. Eum proposed DOB for the robust tension control of the R2R system [11].

Unlike previous studies, we propose feedforward (FF) controller and signal processing technology using the Kalman filter (KF) to improve tension control performance. The web is continuously subjected to various processes for the final production. Particularly, it is important to control the tension when magnitude of the tension varies during process. To improve tracking performance in the transient section where tension changes, the feedforward controller was applied [12–14]. Moreover, a high-price load cell with a highly accurate and noiseless signal should be used to improve tension control performance in the R2R system. In order to reduce costs, which is an unavoidable issue in today's industry, the KF algorithm was suggested for signal processing [15–18]. Almost all load cell applications require filtering to remove noise from the measured signal. A signal filter has a trade-off between noise suppression and phase delay. Phase delay may cause a change in phase margin, stability of the whole system, and accuracy of present information in the signal. Unlike the low pass filter (LPF), a conventional signal filter, the KF minimizes the phase delay [19]. The KF is capable of processing the signal by overcoming the phase delay problem even using a low-cost load cell.

This paper suggests a robust and precise tension control method for the R2R system and is organized as follows. In Section 2, we suggest dynamic model of the R2R system and how to identify the nominal parameters of the system experimentally. In Section 3, the design of a signal filter is presented. In Section 4, the proposed tension control for the R2R system is introduced and robust stability is shown using the small-gain theorem [20–22]. The performance of the proposed control method is verified with experimental results in Section 5. Finally, Section 6 provides some concluding remarks.

2. System Modeling

In this section, modeling of the suggested R2R system is introduced. Moreover, we suggest how to identify the nominal parameters of the system through an experiment [23].

2.1. Dynamic System Model

As shown in Figure 1a, the web is fed at a constant speed by the feeder motor and is wound with the proper tension by the rewinder motor. The web dynamics Equation (1) and roll-to-roll system dynamics model Equations (2)–(8) used throughout the paper are described in References [11] and [24]. The web can be represented in terms of spring and dashpot elements, which are described by the elasticity modulus E and the viscosity modulus η. Lumped parameter models are expressed as the Kelvin-Voigt model for web dynamics. In the case of the Kelvin-Voigt model, the web tensile stress σ is expressed as follows:

$$\sigma(t) = E\varepsilon(t) + \eta\frac{d\varepsilon(t)}{dt} \tag{1}$$

where ε is web strain. Then web tension force f is related to the web tensile stress σ and web cross sectional area A as follows:

$$f(t) = A\sigma(t) \tag{2}$$

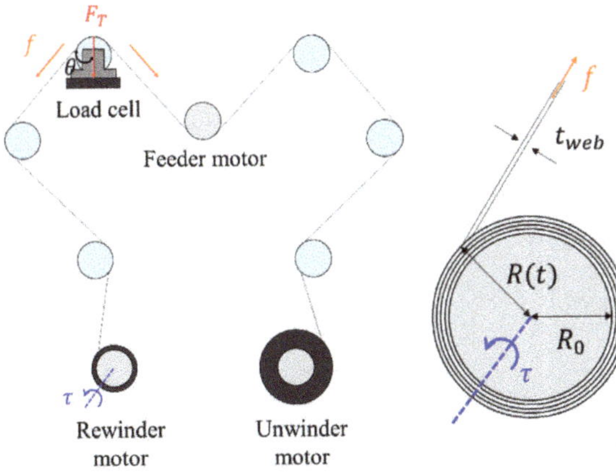

Figure 1. (**a**) Schematic of a proposed roll-to-roll (R2R system); (**b**) Schematic of a rewinder system.

The load cell signal output F_T varies depending on the web tension force f. According to Figure 1a,b, it is expressed as follows:

$$F_T(t) = 2f(t)cos\theta + F_d(t) \tag{3}$$

where F_d is external disturbance. Here, the load cell signal F_T feeds back to control the web tension force f. The torque τ generated by the rewinder motor is related to the web tension force f as follows:

$$f(t) = \frac{\tau(t)}{R(t)} \tag{4}$$

$$R(t) = R_0 + \frac{t_{web}}{2\pi}\theta_m(t) \tag{5}$$

$$F_T(t) = 2\tau(t)cos\theta\left\{R_0 + \frac{t_{web}}{2\pi}\theta_m(t)\right\}^{-1} + F_d(t) \tag{6}$$

where R is the radius of the web, R_0 is the initial radius of the web, t_{web} is the web thickness, and θ_m is the angular position of the rewinder motor. The transfer function of the system from torque input τ to load cell signal output F_T is calculated as follows:

$$P(s) = \frac{F_T(s)}{\tau(s)} = 2cos\theta\{R_0 + \frac{t_{web}}{2\pi}\theta_m(s)\}^{-1} \tag{7}$$

where P is the actual plant. Finally, we designed the plant as a first order lag element because it is a simple structure and represents a system with time lag. It is expressed as follows:

$$P(s) = \frac{F_T(s)}{\tau(s)} = \frac{2cos\theta\{R_0 + \frac{t_{web}}{2\pi}\theta_m(s)\}^{-1}}{1+\tau_n s} \tag{8}$$

where τ_n is a time constant. The nominal model was obtained using an experimental method known as system identification, which develops mathematical models of dynamic systems from the measured input and output data of the system. The process of accurately obtaining a nominal model is important. Because the proposed KF in Section 3 and the tension controller, FF, and DOB in Section 4 largely vary in performance, depending on how accurately the nominal model is obtained, the experiment was carried out in three cases depending on the state of the film being wound. When step input with various magnitudes is applied to the system, the dynamic model is determined through the output data measured by the load cell. As shown in Figure 2, we focused on the characteristics of the system by 100 N because the experiment was performed from 0 N to 100 N.

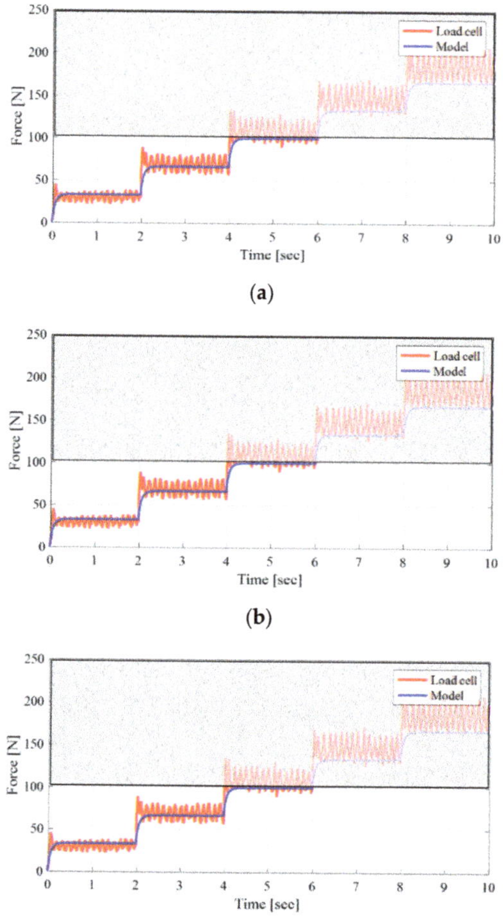

(**a**)

(**b**)

(**c**)

Figure 2. Experimental results of system identification: (**a**) Empty film; (**b**) Half film; (**c**) Full film.

3. Design of a Signal Filter

Signal processing of the sensor using a filter is essential to achieve precise control. Filters have a trade-off between noise reduction and phase response, so the more noise is reduced, the higher phase delay. The phase delay has an adverse effect on the phase margin and stability of the system. In this paper, it was possible not only to eliminate the noise contained in the observed data but also to minimize the phase delay by using the KF. The KF is an optimal estimation method that finds the true value of variables from a set of noisy measurements. In this section, we introduce how to design the KF and signal processing results. The design procedure of the KF is as follows. Firstly, the state space model of the linear system should be constructed. Subsequently, the state variables of the system are estimated by the KF algorithm using the state space model and measured values of the target system.

The KF algorithm has two steps: prediction process and estimation process. Figure 3 illustrates the relationship of the KF with the actual system [15].

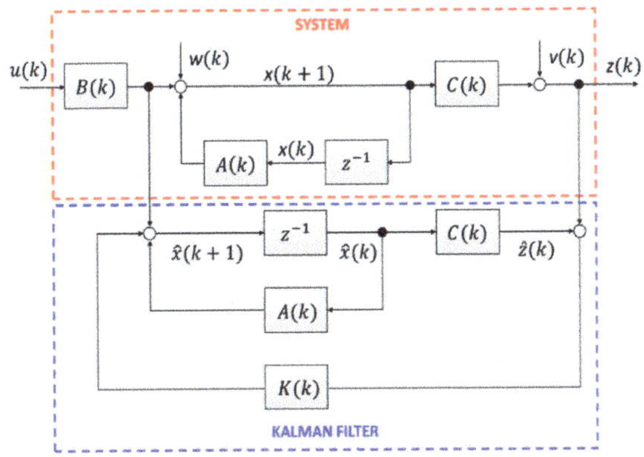

Figure 3. Kalman filter (KF) structure.

3.1. *Linear State Space Model*

To design a KF, the linear state space model must be first obtained by modeling the target system. The performance of the KF depends largely on how similar the system model is to the actual system. Equation (8) can be described by the following differential equation:

$$\tau_n \dot{F_T} + F_T - 2cos\theta R_n{}^{-1}\tau_e - F_\xi = 0 \tag{9}$$

where R_n is the nominal radius of the web and F_ξ is the sensor noise. Defining the state variables $x(t) = F_T(t)$, $w(t) = F_\xi(t)$, the state differential equation is

$$\dot{x(t)} = -\frac{1}{\tau_n}x(t) + \frac{2cos\theta R_n(t)^{-1}}{\tau_n}u(t) + w(t) \tag{10}$$

$$z(t) = x(t) + v(t) \tag{11}$$

where x is the state of the system, z is the output of the system, u is the system input, w is the system noise, and v is the measurement noise.

3.2. *Kalman Filter Algorithm*

The KF algorithm is divided into two stages: prediction process and correction process. Based on the system model variables A and Q, the prediction process guesses how the estimated value $\hat{x}_k$ changes when the time changes from t_k to t_{k+1}. A is the system matrix and Q is the covariance matrix of W. Depending on how the system model variables, such as A, Q, R, and H, are selected, the performance of KF varies. Equations (12)–(17) refer to the KF algorithm mentioned in Reference [15]. The following equation is the state variable and error covariance, which are the predicted variables as follows:

$$\hat{x}_k{}^- = A\hat{x}_{k-1} + Bu_k \tag{12}$$

$$P_k{}^- = AP_{k-1}A^T + Q \tag{13}$$

In the correction process, the KF's final estimated value $\hat{x}_k$ is calculated using the system model variables H and R. H is the measurement matrix and R is the covariance matrix of V. The KF computes the final estimated value $\hat{x}_k$ by adding the predicted value $\hat{x}_k^-$ and the current measurement Z_k multiplied by the appropriate weight value as follows:

$$\hat{x}_k = \hat{x}_k^- + K_k(Z_k - H\hat{x}_k^-) \tag{14}$$

Here, the weight value K_k is known as a Kalman gain, and is newly calculated by repeating the algorithm differently from the LPF and high pass filter (HPF) as follows:

$$K_k = P_k^- H^T \left(HP_k^- H^T + R\right)^{-1} \tag{15}$$

P_k is the error covariance. The error covariance reflects how far the estimated value is from the true value as follows:

$$P_k = (I - K_k H)P_k^- \tag{16}$$

According to the definition of error covariance, if P_k is large, the estimation error is large, and if P_k is small, the estimation error is then small. Error covariance is defined as follows:

$$P_k = E\{(x_k - \hat{x}_k)(x_k - \hat{x}_k)^T\} \tag{17}$$

where $E(\cdot)$ refers to the expected value.

Noise covariance matrices Q and R have limitations to be determined analytically because it is difficult to grasp noise characteristics with multiple errors. In this paper, the Q and R values were determined experimentally through trial and error process. The larger the Q, the more influenced by the measured value. Since R has the opposite tendency to Q, the larger the R, the less affected the measured value.

3.3. Signal Processing Results

Figure 4 shows the load cell signal, the signal processed by the LPF, and the signal processed by the KF. Unlike LPF, which has a limitation in lowering the cutoff frequency due to the phase delay, the KF further reduced noise.

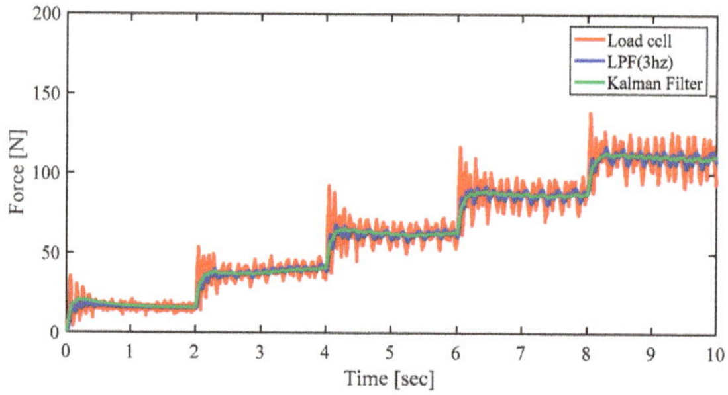

Figure 4. Results of signal processing.

4. Control Design

In this study, the main purpose was to control the tension robustly and precisely in all sections with constant and varying tension. To achieve this goal, we propose a tension controller design based on a model expression and FF controller with excellent tracking performance in the transient section. Moreover, model uncertainty and external disturbance were compensated for by using the DOB. Finally, robust stability of proposed control system was proved by using the small-gain theorem. The proposed tension control structure is shown in Figure 5 and the unwinder motor was also controlled by the same control structure.

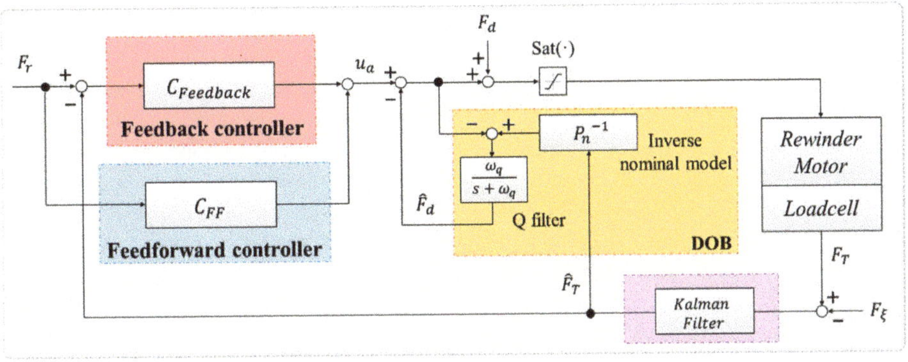

Figure 5. Structure of the proposed controller for R2R system.

4.1. Tension Controller

In this study, the tension controller was designed using a proportional-integral (PI) controller based on the model expression. The model-based PI controller is widely used in practice because it is simple to apply and easy to analyze in the frequency domain [25]. If the zero of a PI controller is designed to cancel the pole of system, the closed loop response characteristic can be made to be equal to the first order lag element. The equation is written as follow:

$$\frac{F_T}{F_r} = \frac{C_{Fb}P_n}{1 + C_{Fb}P_n} = \frac{\omega_c}{s + \omega_c} \tag{18}$$

where F_r is the input signal, P_n is the nominal plant, C_{Fb} is the tension controller, and ω_c is the control bandwidth. Therefore, the tension controller has the following form:

$$C_{Fb} = \frac{K_p s + K_i}{s} = \frac{1 + \tau_n s}{2cos\theta R_n(s)^{-1}} \times \frac{\omega_c}{s} \tag{19}$$

As shown in Equation (18), the frequency bandwidth of the controller is given by ω_c. Therefore, it is possible to design a controller without overshoot only by determining the frequency bandwidth of the desired controller.

4.2. Feedforward Controller

Unlike the feedback control, which can only react after the error between the reference signal and the measured system state occurs, a FF controller that does not feed back the signal provides excellent tracking performance in the transient section because the response speed is fast. Thus, the FF is used to quickly implement the desired behavior when we know the dynamics model of the control system. It is designed as shown below:

$$\frac{F_T}{F_r} = \frac{C_{FF}P_n + C_{Fb}P_n}{1 + C_{Fb}P_n} = 1 \tag{20}$$

where C_{FF} is the FF controller. Therefore, the FF controller has the form of

$$C_{FF} = P_n^{-1} \tag{21}$$

Theoretically, to achieve perfect control without any error between the reference signal and the actual system state, the transfer function must be 1. Therefore, the FF controller must have an exact reciprocal relationship with the plant. In this way, the FF controller is simple in design and powerful in performance, but it has the disadvantage that it is impossible to compensate the uncertainty of the model and the external disturbance.

4.3. Disturbance Observer

The DOB is an effective method of robust control to solve external disturbance and model uncertainty, and is widely used in various fields. It is not only excellent in control performance but also convenient to apply because of its simple structure. In particular, by simply adding the DOB to the inner loop of the existing controller, the performance of the outer loop controller can be guaranteed without separately considering the performance degradation due to disturbance and model uncertainty. The DOB estimates the disturbance, and the estimated disturbance signal is used as a disturbance cancellation input. The DOB is composed of an inverse nominal plant $P_n(s)^{-1}$ and LPF $Q(s)$ that cuts off the disturbance in the low-frequency region. $Q(s)$ is expressed as follows:

$$Q(s) = \frac{\omega_q}{s + \omega_q} \tag{22}$$

where ω_q is the cutoff frequency. To improve the performance in suppressing disturbance, the cutoff frequency should be increased. However, this is restricted to guarantee robust stability, as more unnecessary frequencies can be passed. Thus, it is critical to design an appropriate cutoff frequency.

The effectiveness of the DOB is clearly explained by transfer functions written as follows:

$$\frac{F_T}{u_a} = \frac{P(s)P_n(s)}{Q(s)[P(s) - P_n(s)] + P_n(s)} \tag{23}$$

$$\frac{F_T}{F_d} = \frac{P(s)P_n(s)[1 - Q(s)]}{Q(s)[P(s) - P_n(s)] + P_n(s)} \tag{24}$$

$$\frac{F_T}{\xi} = \frac{P(s)Q(s)}{Q(s)[P(s) - P_n(s)] + P_n(s)} \tag{25}$$

where u_a is the control input and ξ is the measurement noise.

$Q(s)$ plays an important role in compensating disturbance and model uncertainty in the DOB. If the input frequency is smaller than the bandwidth of the $Q(s)$ (i.e., $Q = 1$), the first expression is P_n and the second expression is zero. This means that model uncertainty is compensated for and low frequency disturbances are rejected. On the other hand, if the input frequency is higher than the band-width of the $Q(s)$ (i.e., $Q = 0$), the third equation is zero. Therefore, the high-frequency noise is removed.

4.4. Robust Stabilty Analysis of a Closed Loop Control System

In this section, we demonstrate the robust stability of the closed loop control system. The R2R system has model uncertainty because the inertia of the web in roll form changes during the process. To reflect this model variation, the actual plant is expressed as follows:

$$P(s) = P_n(s)[1 + \Delta(s)] \tag{26}$$

where Δ is the multiplicative model uncertainty.

In Equation (8), the actual plant can be represented by the following parameters as follows:

$$P(s) = \frac{2cos\theta R(s)^{-1}}{1 + \tau_n s} = \frac{1}{As + B} \tag{27}$$

Actually, these parameters have confidence intervals, not single values (i.e., confidence interval A = (0.005, 0.0015), B = (0.15, 0.25)) Moreover, the multiplicative model uncertainty $\Delta(s)$ is calculated as follows:

$$\Delta(s) = \frac{P(s) - P_n(s)}{P_n(s)} \tag{28}$$

The small-gain theorem was applied to demonstrate the robust stability of the closed loop control system. As shown in Figure 6b, the following conditions are given by the small-gain theorem:

$$\left|\Delta(j\omega) \times T\,(j\omega)\right| < 1 \tag{29}$$

where $T(j\omega)$ is the complementary sensitivity function and is expressed as follows:

$$T\,(j\omega) = \frac{P_n(j\omega)C_{Fb}(j\omega) + Q(j\omega)}{1 + P_n(j\omega)C_{Fb}(j\omega)} \tag{30}$$

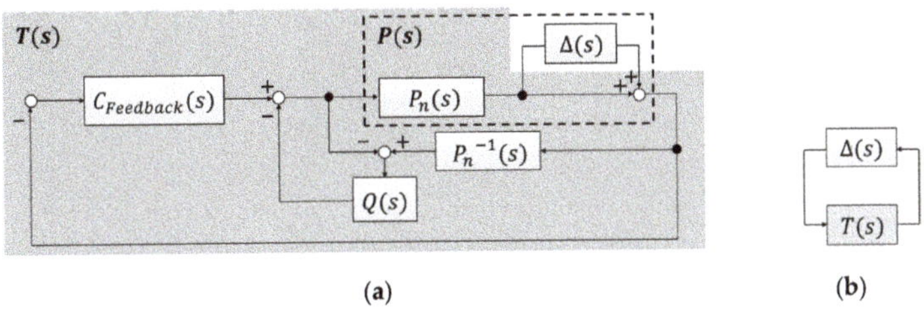

Figure 6. Block diagram of proposed control system: (**a**) Block diagram of the closed-loop control system with multiplicative model uncertainty; (**b**) Equivalent block diagram.

In the bode plot of Figure 7, the magnitude of the complementary sensitivity function $T(s)$ is below the magnitude of inverse multiplicative uncertainty. This means it is possible to implement good tracking performance over the whole range of frequencies.

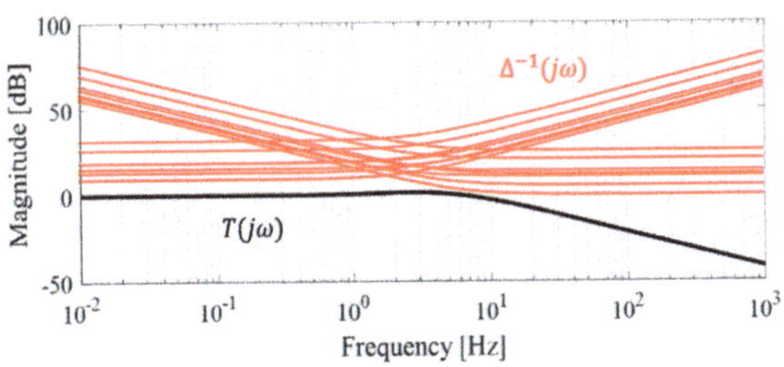

Figure 7. Frequency magnitude response for small-gain theorem check.

5. Experimental Verification

In this section, we introduce the R2R system used in the experiment and the experimental results of the proposed control scheme.

5.1. Experimental Setup

All experiments were conducted on the R2R system depicted in Figure 8 and the procedure was as follows. After fixing the web of the roll form to the rewinder motor, the load cell signal, which varies according to the rewinder motor torque, was controlled in real time by the proposed control algorithm. Then, the opposite ends of the web were fixed to the unwinder motor and the motor was controlled equally. Finally, the feeder motor, which operates only for web feed, was controlled at a constant speed by the proportional-integral controller. A Mitsubishi AC servo motor HF-KP43 and Mitsubishi motor driver MR-J3-70A were driven as rewinder, unwinder, and feeder motors. These were controlled in real time by a Power PMAC ACC 24E3 axis-interface board with 0.1-ms sampling time. The Power PMAC's Integrated Development Environment (IDE) program includes the GNU compiler; thus, we designed the control algorithm with a C-based user code. Figures 8 and 9 show the overall control system and program flow chart for the R2R system.

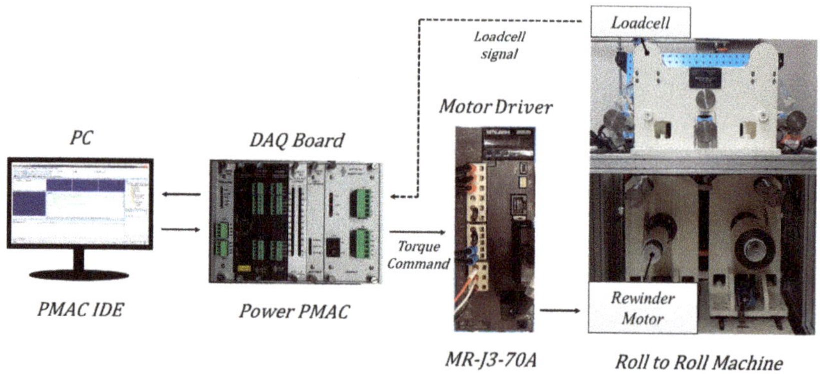

Figure 8. Overall control system for R2R system.

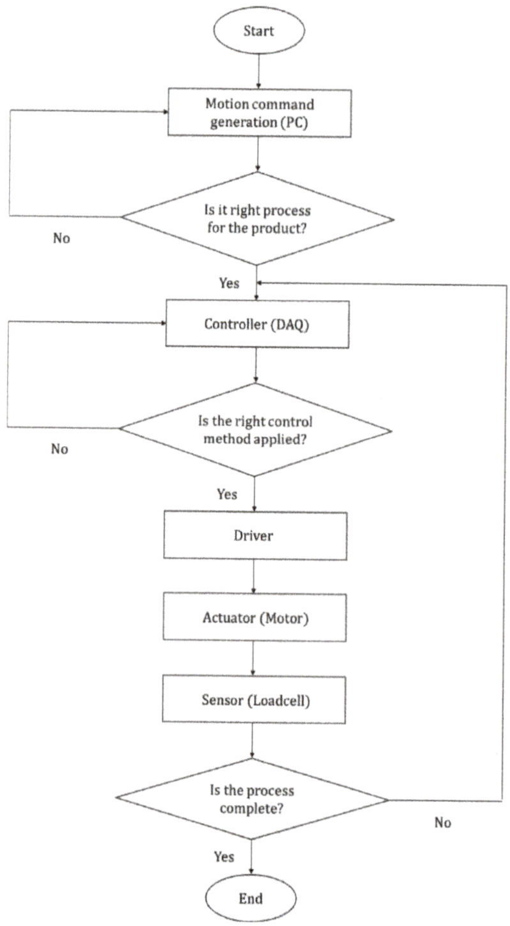

Figure 9. Overall program flow chart for R2R system.

5.2. Experimental Results

In order to evaluate the tension tracking performance of the proposed control scheme, experiments were carried out under the following conditions. The web was transported at a constant speed by the feeder motor and the tension command had a section varying from 0 N to 100 N. Figure 10 shows the experimental results for verifying the performance of the KF. As shown in Figure 10b, the tracking error could be reduced and the bandwidth could be increased when signal processing was performed with the proposed KF relative to the LPF, which is a conventional signal processing method in R2R systems.

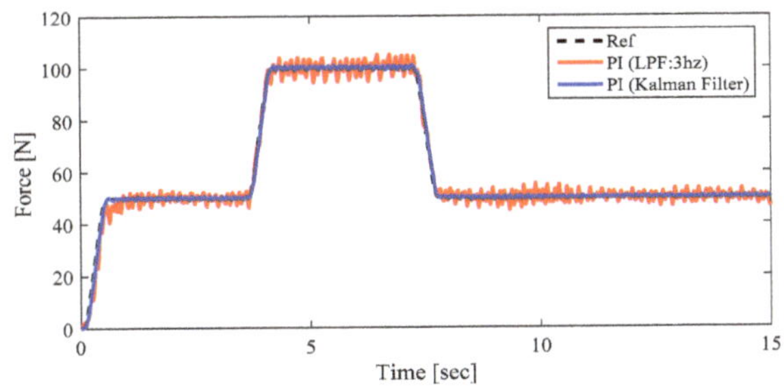

(a)

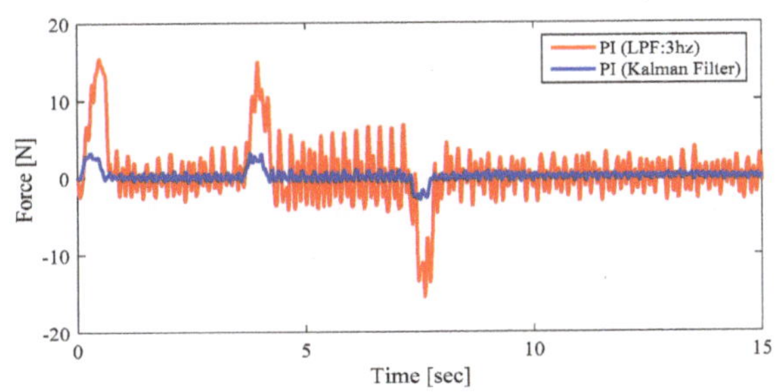

(b)

Figure 10. Experimental results to verify the effectiveness of KF: (**a**) Reference and measured tension; (**b**) Tracking errors.

Figure 11 shows the results of applying the proposed control scheme in the case of signal processing with KF. Particularly, the tracking performance of the FF controller was powerful in the transient section where the tension changes. Moreover, the experimental results demonstrate that the DOB is effective in compensating for disturbance and model uncertainty.

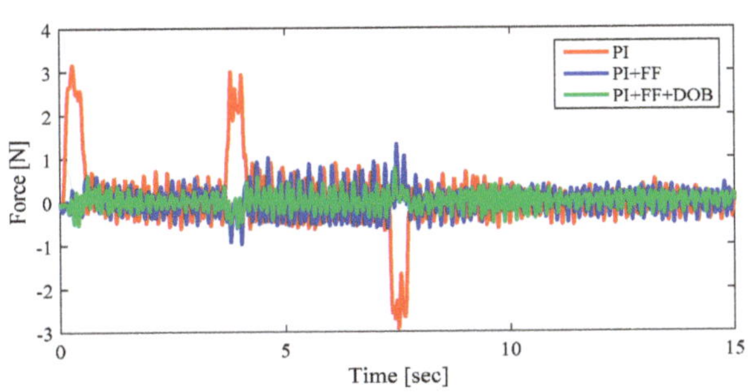

Figure 11. Tracking error for three different control cases.

6. Conclusions

This paper aims to propose a robust and precise control scheme for a R2R machine to overcome the problems of conventional tension control method. A robust control method such as a DOB is essential in R2R systems with model variation and external disturbance. Therefore, a DOB with PI controller was employed to nominalizes the plant and reject external disturbance. The robust stability of the closed loop control system was proved using the small-gain theorem. Moreover, a FF controller was applied to minimize the tracking error in the transient section. The signal processing of the sensor is very important for precise control in the R2R system. The KF not only reduces sensor noise but also overcomes phase delay problem to improve control performance. Experimental results demonstrate the performance of the proposed control schemes.

Author Contributions: H.H. and J.L. have contributed to this paper, performed the experiments, and proposed R2R control algorithms. K.N. and S.E. contributed effort in conducting the experiment and organizing the manuscript as corresponding authors.

Funding: This work has been supported by the Yeungnam University Research Grant 217A580015.

Conflicts of Interest: The authors declare no conflict of interest.

References

1. Søndergaard, R.R.; Hösel, M.; Krebs, F.C. Roll-to-roll fabrication of large area functional organic materials. *J. Polym. Sci. B Polym. Phys.* **2013**, *51*, 16–34. [CrossRef]
2. Ebler, N.A.; Arnason, R.; Michaelis, G.; D'Sa, N. Tension control: Dancer rolls or load cells. *IEEE Trans. Ind. Appl.* **1993**, *29*, 727–739. [CrossRef]
3. Okada, K.; Sakamoto, T. An adaptive fuzzy control for web tension control system. *IEEE Trans. Ind. Appl.* **1998**, *34*, 1762–1767.
4. Koc, H.; Knittle, D.; de Mathelin, M.; Abba, G. Modeling and robust control of winding systems for elastic webs. *IEEE Trans. Control Syst. Technol.* **2002**, *10*, 197–208. [CrossRef]
5. Shin, K.H.; Kwon, S.O. The effect of tension on the lateral dynamics and control of a moving web. *IEEE Trans. Ind. Appl.* **2007**, *43*, 403–411. [CrossRef]
6. Komada, S.; Ishida, M.; Ohnishi, K.; Hori, T. Disturbance observer-based motion control of direct drive motors. *IEEE Trans. Energy Convers.* **1991**, *6*, 553–559. [CrossRef]
7. Sariyildiz, E.; Ohnishi, K. Stability and robustness of disturbance-observer-based motion control systems. *IEEE Trans. Ind. Electron.* **2015**, *62*, 414–422. [CrossRef]
8. Kim, B.K.; Chung, W.K. Advanced disturbance observer design for mechanical positioning systems. *IEEE Trans. Ind. Electron.* **2003**, *50*, 1207–1216.
9. Lee, H.S.; Tomizuka, M. Robust motion controller design for high-accuracy positioning systems. *IEEE Trans. Ind. Electron.* **1996**, *43*, 48–55.
10. Bhattacharyya, S.P.; Chapellat, H.; Keel, L.H. *Robust Control: The Parametric Approach*; Prentice-Hall: Upper Saddle River, NJ, USA, 1995.
11. Eum, S.; Lee, J.; Nam, K. Robust tension control of roll to roll winding equipment based on a disturbance observer. In Proceedings of the IECON 2016-42nd Annual Conference, Florence, Italy, 23–26 October 2016; pp. 625–630.
12. Jamaludin, Z.; van Brussel, H.; Swevers, J. Friction Compensation of an XY Feed Table Using Friction-Model-Based Feedforward and an Inverse-Model-Based Disturbance Observer. *IEEE Trans. Ind. Electron.* **2009**, *56*, 3848–3853. [CrossRef]
13. Shin, K.H.; Kwon, S.O.; Song, S.H. Feedforward control of the lateral position of a moving web using system identification. *IEEE Trans. Ind. Appl.* **2004**, *40*, 1637–1643. [CrossRef]
14. Boerlage, M.; Steinbuch, M.; Lambrechts, P.; van de Wal, M. Model-based feedforward for motion systems. In Proceedings of the 2003 IEEE Conference on Control Applications, Istanbul, Turkey, 25 June 2003; Volume 1, pp. 1158–1163.

15. Halimic, M.; Balachandran, W. Kalman filter for dynamic weighing system. In Proceedings of the IEEE International Symposium on Industrial Electronics (ISIE), Dubrovnik, Croatia, 10–14 July 1995; Volume 2, pp. 786–791.
16. Ljung, L. Asymptotic behavior of the extended Kalman filter as a parameter estimator for linear systems. *IEEE Trans. Autom. Control* **1979**, *24*, 36–50. [CrossRef]
17. Girgis, A.A.; Hwanh, T.L.D. Optimal Estimation of Voltage Phasors and Frequency Deviation Using Linear and Non-linear Kalman Filtering: Theory and Limitations. *IEEE Trans. Power Appar. Syst.* **1984**, *103*, 2943–2951. [CrossRef]
18. Dogariu, L.-M.; Ciochină, S.; Paleologu, C.; Benesty, J. A Connection between the Kalman Filter and an Optimized LMS Algorithm for Bilinear Forms. *Algorithms* **2018**, *11*, 211. [CrossRef]
19. Shaik, S.; Popat, J.; Kumar, T.K. Kalman filter based phase delay reduction technique. In Proceedings of the International Conference on Recent Trends in Information Technology (ICRTIT), Chennai, India, 8–9 April 2016.
20. Lanzon, A.; Petersen, I.R. Stability robustness of a feedback interconnection of systems with negative imaginary frequency response. *IEEE Trans. Autom. Control* **2008**, *53*, 1042–1046. [CrossRef]
21. Nam, K.; Oh, S.; Hori, Y. Robust yaw stability control for electric vehicles based on active steering control. In Proceedings of the IEEE Vehicle Power and Propulsion Conference, Lille, France, 1–3 September 2010; pp. 1–5.
22. Nam, K.; Fujimoto, H.; Hori, Y. Advanced motion control of electric vehicles based on robust lateral tire force control via active front steering. *IEEE/ASME Trans. Mechatronic* **2014**, *19*, 289–299. [CrossRef]
23. Nguyen, V.G.; Guo, X.; Zhang, C.; Tran, X.K. Parameter Estimation, Robust Controller Design and Performance Analysis for an Electric Power Steering System. *Algorithms* **2019**, *12*, 57. [CrossRef]
24. Sakamoto, T.; Fujino, Y. Modeling and analysis of a web tension control system. In Proceedings of the IEEE International Symposium on Industrial Electronics, Dubrovnik, Croatia, 10–14 July 1995; Volume 1, pp. 358–362.
25. Chen, P.; He, Z.; Chen, C.; Xu, J. Control Strategy of Speed Servo Systems Based on Deep Reinforcement Learning. *Algorithms* **2018**, *11*, 65. [CrossRef]

MDPI AG
Grosspeteranlage 5
4052 Basel
Switzerland
Tel.: +41 61 683 77 34

Algorithms Editorial Office
E-mail: algorithms@mdpi.com
www.mdpi.com/journal/algorithms

www.ingramcontent.com/pod-product-compliance
Lightning Source LLC
Chambersburg PA
CBHW061606210926
51850CB00009B/2225
* 9 7 8 3 7 2 5 8 4 8 1 5 7 *